Lean is a powerful discip pal, complex in execution. T. tures the barriers to and enablers of sustainable LEAN results, and those results are ALWAYS worth the effort.

—*Dale Pilger, Former CEO of Truck Bodies and Equipment International, VP Sales & Marketing Asia Pacific Federal Mogul Shanghai.*

Rob, is one of those LEAN practitioners that understands and has proven there is no one way to implement. He is one of the few that have implemented LEAN in several different environments and has adapted and succeeded where others have failed.

—*Dan Bradley, President, Designer Brand Group, Furniture Brands International*

It's refreshing to read a book on lean manufacturing where the author has actually lived through the process. Jablonski's book exposes the complexity that is inherent in converting from a mass production culture to lean manufacturing and points a bright light into the dark corners of this issue. The lessons here apply beyond the factory floor and will help others implement lean across the enterprise.

—*Robert DeAnna, Innovation Creation Manager, Global Fortune 50 Company*

After reading volumes on lean manufacturing, Rob Jablonski brings clarity and purpose to achieving true measurable benefit in our manufacturing operation. This book is a must read that I'll require our "lean team" to read and apply fully into our operations.

—*Ron Safford, President and Owner,*
Parrett Manufacturing, Inc.

The Lean Hangover provides the business leader a practical guide to understanding the nuances and pitfalls of lean manufacturing. It exposes the critical thinking, key learning's and organizational requirements for a successful lean initiative. This is a must read for leading a lean or continuous improvement projects

—*Rich Derian,*
Director Global Logistics & Distribution,
Furniture Values International LLC

Jablonski's book should be mandatory reading for every person involved in a lean transformation. The book helps to bridge the gap between visionary Lean leadership and the practical tools needed to succeed in daily execution. The principles he lays out are the keys to driving real, lasting improvements.

—*David Mullis, General Manager,*
Major Flooring Company

Nothing is as simple as it might seem on the surface. Rob makes the case for a deliberate, thoughtful approach to lean manufacturing, rather than simply jumping off the deep end! If you want to effect positive change in your business while reducing your risk factors, you must read this book!

—*Rich Allen, Executive & Business Coach*

The Lean Hangover provides an understanding of the common pitfalls inherent in implementing lean manufacturing. Jablonski uses his knowledge and experience to provide succinct "lessons" learned on how to successfully implement lean manufacturing around his lean principles. It's a must-read manufacturer's guide to getting the most out of lean.

—*Gene Bryson, Plant Manager, Major Furniture Manufacturer*

The Lean Hangover is a very understandable overview of best practices for corporations that are making the transition to lean manufacturing, or that have found it to be challenging. Jablonski draws upon his experience and key case studies, and his lessons are important ones for reducing risk and adding value to any organization

—*Jonathan Wickert, Iowa State University*

THE
LEAN
HANGOVER

THE
LEAN
HANGOVER

ROB JABLONSKI

Why Businesses Still Struggle
With Lean Manufacturing
& How to Get It Right

TATE PUBLISHING & *Enterprises*

The Lean Hangover
Copyright © 2011 by Rob Jablonski. All rights reserved.

No part of this publication may be reproduced, stored in a retrieval system or transmitted in any way by any means, electronic, mechanical, photocopy, recording or otherwise without the prior permission of the author except as provided by USA copyright law.

This book is designed to provide accurate and authoritative information with regard to the subject matter covered. This information is given with the understanding that neither the author nor Tate Publishing, LLC is engaged in rendering legal, professional advice. Since the details of your situation are fact dependent, you should additionally seek the services of a competent professional.

The opinions expressed by the author are not necessarily those of Tate Publishing, LLC.

Published by Tate Publishing & Enterprises, LLC
127 E. Trade Center Terrace | Mustang, Oklahoma 73064 USA
1.888.361.9233 | www.tatepublishing.com

Tate Publishing is committed to excellence in the publishing industry. The company reflects the philosophy established by the founders, based on Psalm 68:11,
"The Lord gave the word and great was the company of those who published it."

Book design copyright © 2011 by Tate Publishing, LLC. All rights reserved.
Cover design by Kellie Vincent
Interior design by Sarah Kirchen

Published in the United States of America

ISBN: 978-1-61777-110-1
Business & Economics / Industries / Manufacturing
12.03.23

TABLE OF CONTENTS

FOREWORD 11

INTRODUCTION 19

PART 1 UNDERSTANDING THE LEAN MANUFACTURING SYSTEM

MASS PRODUCTION: HOW DID WE GET HERE? 39

MASS PRODUCTION UNDER STRESS 51

ENTER "THE LEAN" SOLUTION 77

LEAN'S PRINCIPLES DEFINED 85

WHY LEAN IS BETTER 91

PART 2 ENABLING THE LEAN SYSTEM TO WORK

LESSON 1: THE LEAN FAB FOUR 107

LESSON 2 LEAN IS CREATING CONTINUOUS
FLOW OUT OF CHAOS 125

LESSON 3: LEAN "MANUFACTURING SYSTEMS
ENGINEERING" CREATES FLOW 139

LESSON 4: THE LEAN MANUFACTURING
SYSTEMS ENGINEERING MODEL 155

LESSON 5: TEN LEAN FLOW TOOLS 171

LESSON 6: BEWARE THE MRP MONSTER
WITH LEAN SHOP FLOOR FLOWS 183

PART 3 AVOIDING BOUNCE BACK
FROM THE LEAN SYSTEM

LESSON 7: LEAN NEW PRODUCT DEVELOPMENT 193

LESSON 8: BUILD-TO-ORDER IS NOT THE GOAL OF ALL LEAN! . . . 209

LESSON 9: LEAN CULTURES DRIVE EXCESS CAPACITY 219

LESSON 10: LEAN AND CI CAN KILL INNOVATION! 229

PART 4 THREE CASE HISTORIES:
THE GOOD, THE BAD AND THE UGLY

LESSON 11: THREE PRODUCT DESIGN STRATEGIES
MAKE TOYOTA A GREAT LEAN MANUFACTURER 245

LESSON 12: MISDIRECTED LEAN NEARLY
KILLED THIS FURNITURE BUSINESS 259

LESSON 13: CABINET BUSINESS:
EVOLVE YOUR LEAN OR DIE 281

AFTERWORD 295

FOREWORD

Only the paranoid, those who are constantly looking over their shoulder to see who is creating something new that will destroy them, will survive.

—Excerpted from *Only the Paranoid Survive*,

Andy Grove, Former Intel Chairman

WHY YOU SHOULD READ THIS BOOK

Your business has to work hard every day to survive. Your competitors are always breathing down your neck, working to take your cheese away from you with newer products at lower costs. Your big box distributors demand innovative new products every year, along with cost reductions for your existing products. And, your customers are always looking for more value delivered through better products at lower costs. This is the fundamental drumbeat of capitalism and it beats endlessly on and on. Surviving in today's fast-paced, global marketplace requires that business leaders develop competitive new *products and processes* continuously; they must develop a company culture that can continuously deliver innovative products at competitive costs.

And, one of *the* most powerful ways businesses have been able to improve their competitiveness over the past two decades has been by moving away from mass production to lean manufacturing. Businesses have implemented lean manufacturing because it makes them more money. That is the long and short reason behind why we've all been chasing this "lean" thing for twenty plus years. When done well, lean manufacturing can help businesses compete in this dog-eat-dog market place with lower production costs, higher quality products, better new product innovations and more reliable delivery lead times.

Lean manufacturing has had such a huge transformational impact over these past twenty years that it has become contagious in the business world, reaching nearly iconic status. As a result of its profit improving impact, lean manufacturing has become ubiquitous. It seems that every manufacturing business surviving today has had some type of lean manufacturing initiative. You can hardly find a manufacturing job posted that doesn't require lean manufacturing, kaizen, six sigma, continuous improvement or the like in experience.

And, many, many businesses have successfully learned to use the lean system to generate stronger profits and growth through their operations. However, and in my experience, there are way too many businesses that are still struggling to get lean right. Even today, even after what feels like a barrage of lean how-to's - training, books, consultants, workshops and such, there are still way too many businesses that aren't getting this lean thing right; some that I've observed, some that I've worked with and unfortunately, some that I've worked for. And, some of these lean wannabe's have even directly damaged their manufacturing performance as result of their misguided efforts to become "Lean". Some have degraded their manufacturing to the point that they've fallen into a downward spiral, creating serious financial stress and even business failures as a result of poorly executed lean changeovers.

Why do so many businesses still struggle today to make the lean manufacturing system work? My answer here may "feel" short and over simplified, yet is has been the key to every lean success and failure I've been involved with. Every lean success that I've experienced and every failure, and all of the stragglers in between, have come down to one thing–Leadership.

Let me explain this further. Every business leader today knows that they want to run their operations leaner, so they chase this prize called lean manufacturing. But the simple failure for those who do not lead their business to a successful lean conversion is that they don't understand how to create and sustain an organization that actually does the work of becoming lean. Lean is not just a different way to run manufacturing within your existing organization. It requires *deliberate* leadership action to develop a new set of organizational skills which can design and implement lean manufacturing solutions throughout the business.

Yes, the penalties of not getting the lean conversion right can be severe. Businesses can and have literally fallen into a downward spiral as a result of their ill-founded lean directions. Businesses that change to continuous flow manufacturing with small batch sizes, before creating excellent process capability and rapid, cost-effective machine setups, can upset their delicate balance of production flow. Pull the wrong levers too early, or under resource a key lean capability and you can literally put the business at risk and do so in fairly short time. Misdirected lean efforts can directly cause throughput to plummet and manufacturing costs to skyrocket, as I have experienced and included in my case history examples.

And, manufacturing leaders who jump into the lean conversion without a clear commitment to developing the key organizational capabilities that make lean successful for the long term can turn lean into a case of darned if you do and darned if you don't: Charge ahead into lean without developing the right organizational skills and you will create a business that sputters with lean, one where you may even upset the delicate balance of throughput in your operations. Or, stick your head in the sand and stay with your old mass production system and you will eventually be outdone by your competitors who get lean better than you do.

These poorly founded lean efforts and results are the source of the title of my book, *The Lean Hangover*. As in life's other hangovers, these miss-guided lean efforts make positive steps early on during the lean changeover, only to be let down by the long term results. Our well intended, but misguided lean efforts yield much less than what we had intended, and in some of the worst cases examples businesses have even ended up in dire financial shape as a direct result of their miss-directed lean initiatives.

Lean Hangovers can take many specific forms, however in my experience they generally fall into two categories; *early lean failures* that keep us from getting the lean system to operate well right from the start, and *later-life lean pitfalls* that creep up consistently, even in successful lean manufacturers.

These first kinds of lean hangovers, the early lean failures, are almost always caused by not committing to and building the right organizational capabilities that are required to support successful lean manufacturing. There is a set of consistent, organizational capabilities that are the key to getting lean successfully started and keeping it moving forward. Not committing to building these core competencies early on as part of the lean transformation will keep you from successfully reaching critical mass, without the right skills to design lean process flow solutions across the business. Without these, your lean implementation sputters and you end up with a lean manufacturing system that limps along instead of the higher performing business you intended.

These early weaknesses can manifest themselves in several ways. In the typical one, businesses start to implement lean but end up with flow imbalances, machine downtime, rework and other breakdowns that pollute their shop floor with waste and product setoffs, much as they did in mass production. Lean manufacturing systems that are setup with continuous flow production, but without each process step performing nearly flawlessly result in manufacturing processes that do not flow consistently. Without an organization that can create high process capability and consistent, short machine changeovers, reliable throughput cannot be achieved.

Further, these poor performing lean manufacturing systems can cause significant pain very quickly. Businesses that don't grasp the lean organizational transformation suffer through new challenges with lumpy output but without the human skill sets to systemically solve their barriers to creating smooth, reliable, lean production flow. Without the right organizational capabilities businesses cannot develop reliable, short lead time, continuous flow solutions for their unique processing challenges. As a result of not developing the right organizational focus and skills, businesses can struggle to take lean beyond their simple shop floor flows. This struggle to get over the lean learning curve has even caused some to whimper back to their simpler, tried and tested mass production methods, a risky decision when you are competing in a lean manufacturing marketplace

Secondly, there are longer term lean hangovers that can creep up on you even when you are deep into your journey, when you think have mastered the concepts of lean manufacturing and think you have the formula right. These latent challenges, or "bounce-back" issues, can be a direct by product of becoming a good lean manufacturer. These are pitfalls which are often hidden by calm waters, but creep into lean businesses that don't continue to adapt their approach as the business changes and grows.

I wrote *The Lean Hangover* to help business leaders succeed with the lean manufacturing transformation by better understanding and leading these organizational changes that accompany successful lean manufacturing. There are specific, systemic, organizational solutions that successfully support the lean manufacturing conversion. As the old saying goes…, our systems are giving us exactly the results they are designed to give us, if you don't like your results, change your systems!

The lessons that make up my book are the short set of critical organizational behaviors that have consistently enabled lean success. *The Lean Hangover* includes proven models for structuring and managing the organization to overcome the early challenges that keep us from getting lean to critical mass, and it also addresses the common pitfalls that can drag down even mature lean businesses. My book provides proven solutions to these lean hangovers in a lessons-learned format, based on case histories and lessons I've learned through years of studying and implementing lean. *The Lean Hangover* explains these proven approaches in a concise, how-to format that every business leader can benefit from, even those who are already making significant lean improvements.

Note: The case histories and examples in this book are a compilation created through my observations working in manufacturing and engineering for the past 25 years. Any resemblance to a particular business or individual is purely coincidental, and is unintended.

INTRODUCTION

There is a better way to do it, find it!
—Thomas Edison

WHY SOME ARE STILL STRUGGLING WITH LEAN

It's Monday morning. You pop out of bed, your stomach churning from the moment you reach consciousness. The entire time you're in the shower all you can think about is your work. Your *to-do* list seems endless; you need to find out why your new product is running over its cost targets; you need to understand why one of your lines is way behind schedule; you need to get your mid-year performance evaluations done; you need to get caught up on the seemingly never-ending list of tasks from human resources, legal, accounting, the CEO and board of directors; you need to stay on top of the perpetual list of requirements from the regulators and certification bodies; and last but not least you need to work down the ever-regenerating list of new e-mails in your inbox, begging for your attention every minute of every day!

You rush through your morning routine as usual. Nobody's up yet at home when you leave. Good, you'll catch up with your family on the weekend. Keys in the ignition and *zoom, zoom*, you are off to the office. Finally you arrive at your desk to get the day going. Settling in, you take care of as many emails as possible while draining your first cup of coffee to get the juices stirring. There we go, e-mails done—*click, click, click*—send, send, send… and good, you've made some progress! Glance at your inbox… ughh, several new emails have been added. Everyone else around the business is trying to get caught up too.

Its 7:05 a.m., you look over your schedule and notice you're booked for most of your normal eleven to twelve-hour day. Your schedule is full right through lunch, just like most days. Good thing you get in early so you can use that time to stay ahead of the power curve. Not much for breakfast today, just that dry bagel you had on the way in to work. You are pumping the coffee to get the gears spinning, and the acid churn from the coffee is already starting to bite.

Remind me, what was it they hired you to do here? Oh, right, you are the head of manufacturing or plant manager or production superintendent or some other important manufacturing leader. Right, and you came here to work on improving manufacturing; to help the business become more efficient, to help it become *leaner*!

Right, that was it. Anyhow, that's what they talked about in the job interview a few years back. But right now you more concerned about pushing your operations to get product out the door. In fact, it seems like you need to push your factories nearly every day to meet production schedules.

Wasn't this business supposed to be better than your last company? It was known for its manufacturing capability and for its lean manufacturing initiatives! You thought your operations were going to run like a well-oiled machine when you signed on - because this business professed it was committed to *lean*!

But, you know that today is going to go just like every other day. The factory has not been running like a well-oiled machine lately and it feels like your list of "abnormalities" is worse than ever. It seems like something always needs special attention in your factory—a machine is broken, a part is in short supply, there is a part that doesn't fit, or there's an engineering change that you need to hustle to implement. And what about your people - do you have enough of the right people to fill the workstations so your lines run today?

Okay, you tell yourself not to worry, that these problems are all solvable. You will get through them one by one. You will make them go away like you've always been able to do, and things will smooth out next week, or next month,… you hope! It had better improve soon, you don't know if you can survive in this rat-race forever. *Gambatte Kudasai!* as they say in Japan—"keep your chin up and put a smile on."

Great! Here you are a big-time manufacturing leader, on the fast track, and living the dream!

You remember back to all of the work you did with those Japanese consultants when you first joined the business. Lean was going to reduce all those wastes of a traditional mass production system: the wastes of overproduction, of waiting, of transporting, of excess inventory, and some others you can't think of right now. Back then, when the lean consultants were helping, you were able to get a lean-learning line setup and running for one of your high volume assembly lines and showed some nice improvements there. That was exciting, but why haven't you been able to bring that into the rest of your operations? In reality, you haven't seen things change much for the better overall; you are still chasing the same problems you had when you started with the company. It's easy to see all of the low-hanging fruit lying around, inventory and rework everywhere, as you walk to your meeting.

Anyhow, no time to feel sorry for yourself, you need to rev up and get things going today. As you are walking to your first meeting you begin to think about long term improvements - "Sure I can work through these daily issues, but how do I find the time to actually work on structurally improving my manufacturing operations? How can I improve our overall process capability to reduce these chaotic abnormalities and make my operations more effective and more manageable every day? Wasn't that what our lean manufacturing initiative was going to help with and how do we make that a priority again…?"

Does this story sound familiar to any of you? For me, it is a pretty accurate depiction of many of the days I've spent in manufacturing. When I first took a role in operations some twenty-five plus years ago, I thought that manufacturing simply made stuff all day long; that operations ran like clockwork. Input A, B, C, sprinkle a little labor on it, and presto, out comes X, Y, and Z. No problem, this manufacturing thing should be a piece of cake! But anyone with a real manufacturing background knows that it doesn't work this easily. My portrayal of the typical morning above is not fiction; it is the world we manufacturing drones live in every day.

This experience with lean manufacturing is also probably familiar to many of you. Ever since the Japanese economic boom of the 1980s, manufacturers have been chasing after Toyota's new and improved manufacturing system, now called *lean manufacturing*. Lean manufacturing was supposed to help manufacturers improve their overall competitiveness by generating faster throughput, lower costs, higher quality, better employee engagement and a host of other system-wide benefits. It seems like just about every manufacturing business that is surviving today has had some type of lean initiative; you can hardly find an operations position on the job market today that doesn't require Toyota Production System, lean manufacturing, kaizen or six-sigma experience. There have been countless books written about lean manufacturing and high-performance lean organizations, and lean consultants are a dime a dozen. Lean has been the improve-all, help-all, fix-all business tool for nearly two decades now and this concept of lean manufacturing has been so heavily promoted as "the way" to fix businesses that l*ean* may have inadvertently become the most overused, and abused, buzzword of our business generation!

So at this point lean manufacturing is obviously not new to anyone anymore. If anything, lean may be bordering on the point of being worn-out news. And there are so many resources available to learn how to make the lean changeover, that the last thing we need is another how-to book on lean manufacturing…. right?

Well then, let's just take a quick temperature check to see how you are doing with your lean manufacturing system:

- Are you able to produce products consistently and deliver them on time?
- Have you been able to reduce lead times across your product lines?
- Have you been able to improve your inventory turns by something significant, like two to four times?
- Have you been able to keep lean going as you innovate with your new products, with smoother launches and shorter lead times?

Fine, if you answered yes to my questions, then you are one of the successful who probably understand the lean manufacturing system well and are making significant results. You may not need this book; Feel free to stop right here and give the book to some lean newbie or maybe you can hawk the book on eBay. Go ahead; you won't hurt my feelings, especially since you've already paid for it…! Or, maybe you can trade the book to someone for one of those Clive Cussler novels. Those Dirk Pitt tales are real page turners, much more entertaining than another book on lean manufacturing.

However, if you are like the rest of us who think they understand lean pretty well, but are still not satisfied with how well the business has been able to execute lean, and want to learn more from the lean successes and failures of others, then please read on.

I've worked in and benchmarked more factories than I'm able to count right now during the past 25 years, learning about lean for my own use, and searching for those key elements that make lean successful over and over again. Some of the businesses I've been exposed to have made it part way through the lean conversion, some most of the way, some all of the way, and some have even gone in and pulled back out of their lean initiatives.

Unfortunately, what I've seen and experienced, even today, is that too many businesses are still not getting lean manufacturing right. I've worked in, supported, and helped improve some of these factories, and also I've walked away from a couple of them because the key decision makers were never going to get it right. A handful of these even seriously damaged their businesses directly as a result of their miss-managed lean conversions. The two case histories in the last section of my book are examples where this actually happened.

Those failures and the struggles of many others to get lean right are the source of the name for of my book, *The Lean Hangover*.

I've witnessed too many businesses—smart businesses, with long-standing, successful brands that still just don't completely get the lean manufacturing system. And I've known really smart, successful business leaders that haven't been able to come to terms with the critical organizational changes that successful lean requires.

On the other side of the coin, I've also participated in highly successful lean reinventions, changes from batch-and-queue manufacturing systems to highly configurable build-to-order systems; businesses that learned to use lean across the enterprise as a competitive weapon; lean successes that enabled the business to grow and gain market share, even in a today's tough economy.

If it feels like this concept of lean manufacturing has been knocked around more than a golf ball in the business world over the past twenty years, that's because it just may have! And if lean is really a better manufacturing method, with all of the resources we've had to get lean right, why haven't *all* businesses been able to figure the lean system out yet? Why haven't we been able to learn it, plug it in, and let it play? Why haven't we all been able to take lead times down, gain consistent product throughput and reduce our costs at the same time? And why haven't we been able to take this system across all of our product lines and make it stick?

So why are businesses still struggling to get lean right, you ask? Well, in my experience these lean challenges have always come down to one central cause—the lack of knowledge based, deliberate leadership action. Let me explain further. Every business leader today wants to run their operations leaner, but many still don't understand how to develop and sustain an organization that actually does the work of becoming lean. They mess it up somewhere between the "I want to be lean" and the "how do *we* actually become lean" parts of the process. Lean is not just a different way to run manufacturing within your existing organization, it requires creating new organizational skills that can design and implement lean manufacturing solutions. Lean leaders need to learn to *lead the development of an organization* that can design, implement and sustain lean processes. And this simple point has been the key to every lean success and failure that I've experienced.

Manufacturing is one of the primary basis on which our businesses compete, and creating competitive manufacturing advantage will always be a challenging, individualized process. As they say in politics, all manufacturing is local. We can define broad characteristics for manufacturing systems, yet there has never and never will be a cookbook recipe for either mass production or lean manufacturing systems. There are certain principles and tools that characterize each system, yet the specifics of what make one company more successful with either mass or lean production are unique.

However, and in this context, the process of designing effective lean manufacturing processes has always been much more difficult to get right compared to the same for mass production. At the core, lean manufacturing is built around the fundamental concept that continuous flow production in very small batch sizes will generate lower overall production costs, shorter deliver lead times and higher quality products. The concept that continuous-flow production can deliver better results is not new; it dates all the way back to the early 1900s, when Henry Ford invented his continuous flow automobile assembly line. Continuous flow production has been a tried and true proven manufacturing advantage for nearly 100 years and is central to both lean manufacturing and mass production systems.

However, the most important difference that is new with lean manufacturing is the process of manufacturing in *continuous flow* through *very small batch sizes* and with *very little work in process and finished goods inventory*. In the days of Henry Ford, and continuing through most of the twentieth century, manufacturing, and especially parts fabrication, was performed in very large batch sizes to generate economies of scale. With Henry Ford's assembly line, large batches of identical automobile parts were fabricated, inventoried, and then queued up for the assembly line to build his low cost Model T. Mass production machines required very long and expensive setups, the time and work required to prepare a machine to do its work. Large batch sizes were necessary to "absorb" these setup costs across a large quantity of parts to reduce the setup cost per piece. However, because of mass production's large batch sizes, we end up with gobs of work in process inventory that would take weeks if not months to move through the production system. Large batches in the mass production system directly resulted in very long production lead times.

Further, the system of mass production has always had a common, base level footprint which applied universally across all businesses. In mass production we set like machines together, in dedicated departments of people, and ran big batches of parts through them, so that we could benefit from focused labor, separation of work, and economies of scale. Mass production was an intuitive way to arrange machines on the shop floor and a better way, in its time, to manage human and capital resources.

Take any reasonably educated individual through a mass production factory with its machine processing centers and assembly lines, and they get it. They quickly understand how and why mass production works. Certainly there were many innovations that made one mass production business more effective than another, but the basics were the same. And, as a result of its simple, intuitive design, businesses readily adopted and succeeded with mass production methods.

Lean manufacturing is not this easy to get right. The ways we create low cost, continuous flow production are not as cookie cutter as the basic rules were in mass production. Let me explain this in some detail. The *fundamental* challenge we take on with lean manufacturing is that we work to design cost-effective processes that continuously flow production from machine to machine, in very short lead times. However, the only way we can create this continuous flow production system is by virtually eliminating all long machine setup times and their associated big batches. By doing so, products can then flow in smaller packets from machine to machine very rapidly, even down to one-piece flows; thereby creating very short lead time production flows. Short manufacturing cycle times are one of the core results of lean manufacturing. And, continuous flow, short lead time processing results in more efficient manufacturing operations and a business that ultimately becomes more competitive and responsive to its customers.

However, manufacturing goods in small batch sizes has historically been cost prohibitive due to these built-in high machine setup costs. Herein lays the essence of the fundamental lean manufacturing challenge: how do we design manufacturing systems that can cost effectively process small batches so that products can move very quickly through the plant, without being saddled with these huge setup costs or wait times? The solution to this is simple, in theory: we have to design cost effective ways to virtually *eliminate setup times*. In mass production we simply arranged long setup time, batch processing machines with other similar machines on the shop floor. However, in lean manufacturing we must develop creative methods to significantly reduce machine setup times so that we can cost effectively manufacture in small batches, even down to one piece flows. That in turn allows us to arrange machines side-by-side in continuous flow cells to create rapid, continuous flow production. Short machine setup times enable cost effective small batches, and these small batches, in turn, allow us to arrange machines into continuous flow, lean production. Therefore, short setup times and small batch sizes are *the* fundamental enablers to lean manufacturing.

And as a result, one of the base level requirements to successful lean manufacturing is developing this internal capability to attack and eliminate long setup-time barriers to continuous flow. In lean manufacturing we must develop unique and innovative ways to significantly reduce machine setup times so that we can cost effectively manufacture in small batch, continuous production.

Further, we create another new challenge when we connect dedicated machines in continuous flow processes and take work in process inventory out of the manufacturing system to create continuous flow, as we do in lean. We purposefully create a manufacturing system that is strung very tightly in lean manufacturing. Mass production was primarily *asynchronous*; each operation was disconnected, working largely independently of upstream and downstream operations. When one operation went down others surrounding it could usually continue to march on. As a result, asynchronous mass production could survive, albeit barely, when it was sloppy and had breakdowns.

Effective lean manufacturing is very *synchronous* manufacturing. Again, lean manufacturing is the process of manufacturing in *continuous flow* through *very small batch sizes* and with *very little work in process inventory*. When we build with continuous flow manufacturing processes, with very little work in process inventory as parts move from machine to machine, manufacturing becomes like a rubber band stretched nearly to its breaking point, every minute of every day. As a result, Lean manufacturing crashes quickly and painfully when it is sloppy. Lean's continuous manufacturing processes screech to halt when any one thing fails anywhere within the links of the continuous flow chain - the rubber band snaps! There's almost no work in process inventory to keep lean's continuous flow processes moving when any one part of the manufacturing chain goes down. And, with little work in process or finished goods inventory, when the rubber band snaps we start to impact throughput and customer service levels very quickly. Lean manufacturing is strung so tightly that we can put our basic throughput at risk if we do not have ultra-high process capability for every step in the manufacturing process. If everything doesn't operate reliably all of the time, synchronized lean manufacturing crashes and burns.

As a result of these new, innate challenges that come along with lean, building a successful lean manufacturing business requires fundamental organizational change to how we design, implement and support our manufacturing systems. Creating this reliable, continuous production flow production requires unique and creative process designs, ones where your product and process engineers must work together to innovatively break down the barriers to small batch production. Ones they cannot buy out of a catalog. Successfully changing to lean manufacturing will require you to develop the internal organizational know how and spirit to "invent" your own reliable, small batch, lean process designs, and invent them for all of your production flows, from your simplest to your most complicated ones, and keep them running nearly 100% of the time.

This is the fundamental challenge that comes with the lean manufacturing conversion, and where the breakdowns occur most often that can cause a lean transformation to fail. The greatest challenge in lean manufacturing isn't with the execution of a well designed lean manufacturing system; it is in the *process of designing* lean manufacturing systems that can be executed well. Failure with lean manufacturing is almost always caused by the lack of leadership direction to empower the process of designing, implementing and maintaining lean manufacturing systems. Lean is not just a different way to run manufacturing within your existing organization. It requires creating new organizational capabilities that can design and implement lean manufacturing solutions. Lean leaders need to learn to *lead* the development of an organization that can design, implement and sustain lean manufacturing processes. Successful lean requires leadership that can create an organization that designs these lean manufacturing solutions!

And, when this organizational evolution doesn't accompany the lean changeover, we usually end up with a poorly performing lean manufacturing system, creating the premise and title for my book, what I call The Lean Hangover. We end up with a poorly performing manufacturing system that doesn't provide the reliable, short lead time, consistent production that we expect and should gain through effective lean manufacturing. Our well intended efforts to improve the business through lean manufacturing end up falling short due to a lack of leadership understanding and driving the key organizational changes we need to create alongside the lean manufacturing conversion.

I wrote my book to help business leaders succeed by better understanding these key organizational behaviors and structures that enable successful lean manufacturing. There are specific, systemic, organizational solutions that businesses have developed to support the lean manufacturing conversion. The lessons that make up my book are the short set of these critical factors that have consistently enabled lean success throughout my career. In The Lean Hangover I provide these key organizational practices that the winners use to succeed with lean and the strugglers are still trying to figure out. The Lean Hangover explains these key organizational behaviors through a lessons learned format that every business leader can use to lead their organization to more successful lean manufacturing.

THE STRUCTURE OF THIS BOOK

I've organized *The Lean Hangover* into four parts:

Part 1, *Understanding the Lean Manufacturing System*, explains how the mass production and lean manufacturing systems came to be, how they differ from each other, and why the lean system of manufacturing is better today. This is an important basis to understanding what we are trying to achieve through the lean manufacturing transformation, so that we can get it right.

In Part 2, *Enabling the Lean System to Work*, you will find my solutions for the necessary organizational skills that design, implement and sustain lean manufacturing processes to get lean over the learning curve and make it successful throughout the business. These organizational models are how businesses get the lean transformation successfully started and keeping it moving forward.

Part 3, *Avoiding Bounce Back from the Lean System*, are lessons that address the longer-term challenges that come about even in successful lean manufacturing businesses. These lessons are directed at keeping your lean system successful for the long haul by avoiding common pitfalls that can cause businesses to stumble directly as an outcome of their lean successes.

Part 4, *Three Case Histories, the Good, the Bad and the Ugly* discuss how Toyota's new product design process has led to an even more efficient lean manufacturing system and how two other businesses misfired with their lean transformations, crippling their businesses as a direct result of poorly executed lean changeovers.

PART 1

UNDERSTANDING THE LEAN MANUFACTURING SYSTEM

MASS PRODUCTION: HOW DID WE GET HERE?

> The highest use of capital is not to make more money, but to make money do more for the betterment of life.
>
> —Henry Ford

Mass Production revolutionized business and industry in the twentieth century and, by doing so, changed the world forever. Our mass production based industrial revolution propelled western civilization forward, delivering an almost incredible progression of low cost, innovative products and technologies to the masses. Mass production enabled businesses to manufacture complex, higher quality products at lower costs—costs that made these products available to ever more and more people. As a result, mass production increased the general level of prosperity in our country and has been a direct contributor to the high standards of living we've come to enjoy today in the western world.

However, in the 1980s, global manufacturers started to take note of a new, more efficient manufacturing system which Toyota pioneered to produce better automobiles. This new manufacturing system, called *lean manufacturing*, was based on the fundamental concept that continuous flow production in small batches sizes can lower overall production costs and better satisfy customers with shorter, more reliable order lead times. Business leaders studied this new system to learn how to apply lean manufacturing to their own businesses and from there the lean manufacturing ball started rolling. Over the past twenty years manufacturers around the world have worked to shift away from mass production to lean manufacturing.

My starting point in this journey to become more effective lean manufacturing leaders is to learn first what exactly mass production is and why it worked so well in the past, but not today. This Chapter and the next two explain how lean and mass production differ from each other, and why the lean system is better today. This is an important starting point to understanding what we are trying to achieve through the lean transformation so we can get it right. By understanding our evolution from mass to lean production and what makes lean a better method, lean leaders can develop effective approaches to move their businesses through the lean transformation.

So let's start with the fundamentals of what exactly mass production is, how it came to be, and why it has recently fallen behind in the manufacturing race. This evolution of the mass production system can best be told through the contributions of three historic manufacturing thought leaders: Eli Whitney, Frederick Winslow Taylor, and Henry Ford.

Eli Whitney is best known for his invention of the cotton gin and how it revolutionized cotton farming, making cotton farms more efficient, and leading to plantation size farming. However, in our lean manufacturing journey, Whitney is more important for how he sought to improving military weapons through the manufacture of standardized rifle components. Whitney envisioned that significant benefits could be gained by developing interchangeable parts for firearms for the United States military. With interchangeable parts, if one mechanism in a weapon failed, a new piece could replace it and the weapon would not have to be discarded.

In 1878, Eli Whitney built ten guns using identical parts and mechanisms. He took them to the US Congress and disassembled them right there on the floor of the Congress. Whitney placed the parts in piles, mixed them up and then reassembled all of the guns in front of Congress. Our legislators were impressed with Whitney's exhibition and supported interchangeable parts as a future standard for all United States military equipment.

This concept of interchangeable parts was a potential game changer, but yet too difficult to put into practice at that point in history. Skilled craftsmen had made the parts by hand and at great costs for the guns that Whitney had shown congress. His firearms used the only manufacturing methods available at that time to make them interchangeable—manual filing and hand shaping. Whitney knew that this concept of interchangeable parts was a key improvement to manufactured goods, but the tools and machines required to cost effectively make interchangeable parts had not yet been developed.

However, Whitney's idea eventually proved feasible and was a critical step toward mass production. His "interchangeable parts" concept was one of the fundamental keys to developing standardized product designs, automated machining, and eventually mass production assembly lines. The development of effective manufacturing processes that could manufacture interchangeable parts was well underway in the late 1800s. The concept of semi-skilled labor using machine tools instead of traditional hand tools advanced over time to eventually enable Whitney's concept of interchangeable parts to take hold.

Our second key historical contributor to mass production was Frederick Winslow Taylor. Taylor's work centered on the methods of planning and structuring work flow. His research and beliefs led to Taylor's landmark publication, *The Principles of Scientific Management* in 1911. In his book, Taylor's scientific management methods called for optimizing the way tasks were performed and simplifying the tasks so that workers could be trained to perform their specialized sequence of tasks and motions in one "best" way. Work is to be planned through time studies and the division and balancing of labor content to increase productivity in a manufacturing environment. Before Taylor's methods became widely practiced, most factory workers were highly trained and skilled apprentices, often capable of machining and manufacturing an entire assembly on their own.

After years of experimentation to improve productivity through optimal work methods, Taylor put forward these four principles of scientific management in his book:

1. Replace rule-of-thumb work with methods based on a scientific study of the tasks.
2. Scientifically select, train, and develop each worker rather than passively leading them to train themselves.
3. Cooperate with the workers to ensure that scientifically developed methods are being followed.
4. Divide the work nearly equally between managers and workers so that managers apply scientific principles to planning the work and the workers actually perform the tasks.

If this sounds a lot like today's industrial engineering to you, it is very much so. Taylor is credited with much of the groundwork that led to the standardization of work that could be used to improve productivity, and thus he is generally known as the grandfather of the industrial engineering discipline. So you can credit your beloved Standardized Operating Procedures (SOPs) ultimately to him.

So then, how were Taylor's scientific principles a key enabler to what became known as mass production? What Taylor did was to create a shop floor management methodology and discipline that could be used to control production flow and improve process efficiency. Taylor's Scientific Principles was a philosophy for designing efficient labor methods that matched and complemented production part flow. And, in a "what came first, the chicken or egg" type of discussion, the effectiveness of Taylor's planned production work methods had a direct dependency on Whitney's concept of standardized parts. Think of it this way, standardized work had to be combined with standardized, reliable part flows to be effective. You cannot create and control standardized labor flow on the shop floor if your parts don't reliably move through the shop floor from one operation to the next. That constitutes a job shop where individuals perform labor on an unplanned, as-needed basis due to the lack of reliable, repeatable standards.

Whitney's interchangeable parts concept was the concept that could be married to Taylor's standardized work to make the entire production process more effective. The two concepts were complementary, in a one-plus-one-makes-three kind of synergy. Standardized work combined with interchangeable parts and assemblies enabled manufacturers to build products more efficiently and with higher quality than ever before. In this way, interchangeable part designs and efficient labor management techniques became two of the key concepts that ultimately led to mass-production.

Almost paradoxically, Taylor's principles have e̶x̶t̶e̶n̶d̶e̶d̶ beyond mass production and are still a critical part of today's lean manufacturing systems. Standard work is the foundation of every lean production system and sets the basis for ongoing improvement. Taylor himself firmly believed in the need for continuous improvement and was an interesting precursor to the CI world we live in today. Here is a relevant quote from his 1911 book, *The Principles of Scientific Management*:

> And whenever a workman proposes an improvement, it should be the policy of the management to make a careful analysis of the new method, and if necessary conduct a series of experiments to determine accurately the relative merit of the new suggestion and of the old standard. And whenever the new method is found to be markedly superior to the old, it should be adopted as the standard for the whole establishment.

Continuous improvement may just be innate to the human spirit!

At this point in history we were standing on the doorstep of our industrial revolution, but still needed one more major breakthrough to create the complete system of mass production. The man that pulled it all together is no surprise; he is the father of that invention, the assembly line, and he was *Henry Ford*.

Following is a short summary of Henry Ford's development of the assembly line and mass production:

In the early twentieth century, the automobile was a plaything for the rich. Automobiles were expensive, custom-made machines. Ford was determined to build a simple, reliable, and affordable car—a car the average American worker could afford. In 1907, Ford announced his goal for the Ford Motor Company to create "a motor car for the great multitude."

Ford started off by designing the Model T—a simple car that offered no factory options, not even a choice of color. Ford started production of the Model T in 1908, and while it was less expensive than other cars, it was still too expensive for the masses. Ford knew he needed help to lower the production costs of his cars, so he looked to other industries. He and his team found four principles that helped them set their direction—interchangeable parts, continuous flow, division of labor, and optimized work effort.

Ford used Eli Whitney's concept of interchangeable parts in his Model T so that any part would fit any vehicle. This meant Ford had to improve the materials and the machines of that time that made his parts. Ford worked to replace the skilled labor that used to make these parts by hand with lower skilled labor that could run reliable machines once they were setup.

Ford also had been exposed to the meat packing houses of Chicago and conveyors they used there. He decided he needed to copy that model and bring the Model T work to the workers in a continuous flow so that tasks could be completed in sequence. He divided the labor for the Model T into eighty-four distinct steps and worked to balance the workloads from step to step and to assure the process could be executed quickly and reliably with little setup.

Each worker was trained to do just one of Ford's eighty-four steps. Ford asked Frederick Taylor to do time and motion studies to determine the speed the conveyor should proceed at and the motions workers should use to do their work.

It took Ford more than five years to reach his goal and perfect the assembly line. The Model T had a price of $825.00 when it debuted in 1908. Four years later the price had dropped to $575.00, and sales skyrocketed. By 1914, Ford claimed a 48 percent market share of the US auto market.

The Model T kept its same basic design from 1908 until the last unit rolled off the line in 1927, number 15,000,000!

Out of Ford's determination came the assembly line and the eventual success of the US auto industry—two of the most important innovations in the history of manufacturing and the development of modern society.

Henry Ford may have been the best-known user of Taylor's principles. As an interesting aside, Ford was so successful in applying Taylor's principles in his automobile factories, and it achieved such great productivity that families even began to perform their household tasks based upon the result of time and motion studies. Wouldn't that make for interesting dinner table fodder?

Mass production and the great industrial revolution were off and running. Ford's innovation of the assembly line tied these three fundamental and complimentary ideas together in what then became the most successful manufacturing method in human history, mass production:

1. Interchangeable parts (Whitney)
2. Standard and optimized work practices (Taylor)
3. Controlled work flow and pace through the assembly line (Ford)

However, there is one additional factor that was critical to the overall success of early mass production systems. A low-complexity product offering was part and parcel to the successful development of Henry Ford's mass production system. It was imbedded in the story of Ford's invention of the assembly line and revealed again in Ford's ubiquitous quote, which captures the essence of mass production:

> Any customer can have a car painted in any color he wants, so long as it is black.

Henry Ford knew that in order to create highly efficient, consistent shop floor flow he had to limit his automobile to a very narrowly defined product offering. In fact, Ford didn't just limit his offering; he made only one flavor Model T!

Ford knew that product options would add complexity to both his parts making and his assembly line and that this complexity would greatly reduce the overall reliability of his processes. Ford understood that limited features and options would ensure that he could better control parts quality and also reduce inventory to manageable levels. By limiting his offering to one vehicle make, color, and outfits, Ford could better rely on an inventory of a very limited number of parts to keep his assembly line running with peak efficiency. An assembly line is linked together like the links on a chain, and if one of the processes stumbles, the entire line screeches to a halt. Lost time becomes very expensive if we think about the dozens of people and tools that suddenly go idle along with the loss of productive output. Ford needed a limited product offering with few parts to make his assembly line dependable and to achieve his goal of a low cost "motor car for the great multitude."

The catch in this business approach is that Ford had to have a market to push his products to, with his narrow product offering. At that time "affordable automobile" was the primary purchase driver, and Ford had just such a market opportunity! Ford didn't need color or layout or other options to attract customers yet. Personal choice was not Ford's primary market motivator in these early stages of personal transportation, and thus his simple, low cost, black Model T was hugely successful.

However, personal choice would very quickly become a key automobile market motivator, enabled by the success of mass production. The natural progression from simple mass production manufacturing systems to complex production systems running hundreds, if not thousands of product offerings and variations through the shop floor came quickly and ultimately challenged mass production as the best way to manufacture in the 21st century. We will dive deeper into how this progression played out in the next chapter.

So when we study the history of successful mass production, it really was also characterized by a fourth key aspect, which was part of its overall success—*limited product offerings*.

Summarizing, the four key historical elements of successful mass production were:

1. Interchangeable Parts (Whitney)
2. Standard and Optimized work practices (Taylor)
3. Controlled work flow and pace through the assembly line (Ford)
4. Limited/focused product offerings (Ford)

Following these four core concepts, the world's greatest manufacturers developed highly efficient manufacturing operations as the twentieth century zipped along from one innovation to the next. The application of these mass-manufacturing methods to highly planned and efficient assembly lines brought forth products at costs levels never before achieved. This was a period in time when mass production improved nearly all product categories, from clothing to furniture to TV's and electronics and just about all manufactured goods. Our manufacturing capabilities delivered luxury-level conveniences to the masses across the civilized world, and mass production changed the world we live in forever.

So what happened with mass production that caused it to become outdated and replaced by lean manufacturing, this once hugely successful method of building goods and services for the masses? That is the subject of my next chapter.

MASS PRODUCTION UNDER STRESS

> A bad system will defeat a good person every time.
> —Edward Deming

Today we love stuff. We love lots of stuff, and we love great stuff at low prices. We also love choices. We want to be able to buy the products that fit our own personal needs and style. We want the Motorolas, the Nikes, the Apples, the Toyotas, the Polos, the "whatevers" that fit our personal needs, taste, color, size, and price points.

And, naturally, businesses want us to want *their* stuff. They work tirelessly to make their stuff better than the competitor's stuff so that we buy their stuff. This Yin-Yang of consumer need and market fulfillment has created a marketplace today where we have more choices than ever before.

I have to admit that I, for one, am a bit of a bona-fide consumer. I am an engineer at heart, and I love new innovations. I love the technology and the capabilities that new products bring with them. Sign me up for new and better and innovative products as long as we are conscientious not to overdo this consumerism and recycle wherever possible.

So what does all this stuff about stuff have to do with mass production and lean manufacturing? Well, with the wheels of mass production humming throughout the mid-twentieth century, propelling the civilized world ever forward towards higher standards of living, an evolution occurred that changed the effectiveness of mass production, one that would eventually tilt the tables in favor of lean manufacturing as a better way.

What was that change? Simply put, the change was that we became really good at mass production; maybe we just became *too* good at mass production. Manufacturing techniques, machines, processes, and information flow all continued to improve throughout the twentieth century. Our mass manufacturing capabilities progressed to higher and higher levels of efficiency, and that, in turn, enabled us to make more stuff, more diverse stuff, more of what we actually wanted and perceived our customers needed. However all of this diversity and product complexity also started to challenge the fundamental effectiveness of the mass production system as I will explain in this chapter. In a sense, mass production's success ultimately became its own demise.

Capitalism is built on the fundamental need for businesses to grow to continue to attract new investors and to provide a good return to stakeholders. New products and innovation are the lifeblood of our businesses. Profitable growth through new products and innovation increases a business's value, which in turn attracts new investors who give their dollars to the business to invest in even more growth. These are the cranks that turn the gears of capitalism.

So what do successful companies do as they look for new ways to grow their businesses? They bring out more and more stuff; new product offerings, product adjacencies and extensions, more product options, and more geographically dispersed distribution. I feel like we need a "Lions and tigers and bears, oh my!" right here…!

Over the past hundred years or so, companies created new and stronger capabilities through mass manufacturing. And, these businesses also had the ongoing, never-ending need to develop and sell more stuff to make their businesses successful. Mass production created the sheer capability to make more, better stuff, and our free market drove us to do just that.

Take a look, for example, at the early automobile industry of the 1920s, and you will see this playing out already in the early years of mass production. Less than two decades after Ford perfected the assembly line around his singular Model T offering, Alfred Sloan—then president of General Motors—explained in his 1924 annual report to shareholders his now famous marketing strategy of "a car for every purse and purpose". Sloan also created the concept of planned obsolescence at GM through annual styling changes and periodic model revisions. Another of his ideas was the "ladder of success," whereby the five brands at GM—Chevrolet, Pontiac, Oldsmobile, Buick, and Cadillac—corresponded to increasing social status so that customers could stay in the GM family as they progressed through their own personal successes. Sloan's strategies recognized the power of consumer choice and the need for diverse product offerings to win over these consumers, and he bet on those as a better business model for General Motors.

Contrast these strategies back to Henry Ford's 1910s "Any... car painted in any color as long as it is black," and you get the essence of where mass production was taking us, and it was headed there rapidly. In fact, Sloan's diversification concepts, coupled with Ford's resistance to do so, launched GM to industry leadership by the 1930s, a position it held for over seventy years. Following Sloan's strategies, GM became the largest, most successful, and most profitable enterprise the world had ever known, representing 10 percent of the US economy at the height of its success.

This is just one simple example of mass production's rapid evolution toward broader and more complex product offerings, but a similar progression was playing out in other industries as well. Mass manufacturing gave us the fundamental manufacturing capability to develop and expand our product lines to meet evolving consumer wants and needs. This new manufacturing capability enabled us to grow our product portfolios and in many cases grow them until our factories were bulging at the seams with ever more and more complexity to make all of this stuff. Mass production systems had served us well for many decades, yet these more complex, more diverse, and more customized product lines that were ultimately born out of mass production's capabilities began to unleash new challenges on the shop floor, challenges that eventually led to the development of lean manufacturing as a better way. I will explain how this came to be in the following portrayal of a typical mass production system.

MASS PRODUCTION AND "BATCH-AND-QUEUE" MANUFACTURING EXPLAINED

How exactly did product diversification and complexity challenge the shop floor and start to create significant problems in mass production businesses, ultimately tipping the scales toward lean as a better solution? The following is how this played out.

Since mass production was built around interchangeable parts, we setup machines to be able to make these interchangeable parts in a way that was intuitive and cost effective at the time. In mass production, manufacturing machines were generally single-purpose machines that were located with similar machines on the shop floor. These are often referred to as machining centers. Mills were located with other mills, lathes with lathes, presses with presses, and so on. This arrangement was intuitive; similar types of work were accomplished in focused locations within the factory and labor could be controlled and managed in pools of similar kinds of work. This arrangement made it easy to manage part flows and the application of labor to process parts: if we needed to move a worker from one press to run another press, the management, pay scales, and know-how were all easily handled within machining centers.

A typical factory in the days of mass production might look like the diagram below where islands of like machines were arranged together physically. Large stamping presses are arranged in a row, as are the molding machines and the specialized machines in the machining department. It's not shown in the diagram as a detail, but even within departments like machining, similar machines were sub-grouped together; mills with other mills, lathes with lathes, and the like as stated earlier.

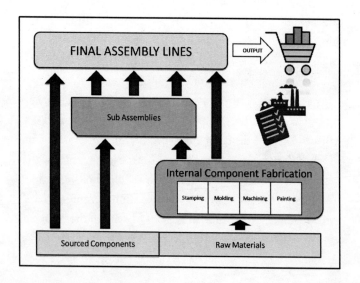

Simply put, the breakdown in mass production was ultimately caused by long travel paths, collisions, and wait times that parts and assemblies incurred while traveling through the mass production shop floor. And the scales were tilted more and more over time in direct relation to the levels of product complexity that we added to our factories.

Let's put this into real terms by thinking like a mass production part for a moment, say a simple bracket as follows:

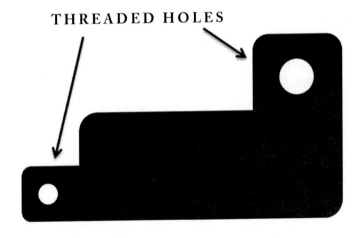

THREADED HOLES

SIMPLE METAL BRACKET

Our bracket will start as raw sheet metal which is moved to the stamping area where the shape is stamped out. Then it might be moved on to one or two other machines to thread the different sized holes and finally tumbled or deburred. Then the bracket is moved over to painting, and finally our part is ready to be queued up in inventory for final assembly.

This is a really simple part flow, but you can still see that the bracket moves around the factory a fair amount. Another observation is that the machining time at any one machine would be very short for this part. The stamping die is likely to be designed to make two or four brackets per hit so we can stamp out these parts at a rate of a few seconds per part.

Let's step back and think about setting up the die in the stamping press for a moment. It wouldn't be unusual for this die setup to take one hour or longer in a typical mass production manufacturing system—removing the previous die, installing the two halves of the new die, aligning everything, and then trying it out and making our quality control checks to ensure the setup is ready for a production run. Even if we are really good at die changes, this could easily take say fifteen to thirty minutes.

So with all of this time invested in setup, combined with the capability to stamp out many parts per minute, we are going to want to make several dozen of these brackets each time we decide to run them. Otherwise our setup costs are too high to absorb over just a few parts. And we need the capacity of this machine running more often than not to be able to make and sell low-cost products. We can't setup for fifteen minutes and just run parts for one or two minutes. Most of our capacity would be wasted in setups, and parts would resultantly cost too much.

This is precisely where the term *batch* came from as in "batch and queue manufacturing". Batch is used to describe how parts move through a mass manufacturing system. We build parts and assemblies in batches in a mass production system to reduce the cost per piece. (See sidebar on "Why We Benefit from Batches in Mass Production.") This is one of the reasons Henry Ford designed his Model T with only one basic design. He could benefit from lower costs of a standardized set of parts because his batches could be made fairly large in size.

Okay, so far this all makes sense. This looks like typical manufacturing work. Making this bracket is no big challenge as long as the machines are available when we need them; with available machines this part will move through the factory like a sled on fresh snow. So if I am Henry Ford, and I only have a few brackets like these to make, no problem; my business model works great. If we have only three or five or maybe even ten different styles of these parts in our factory, mass production is very manageable. The parts will move through the machines quickly and without long wait times.

However, mass production really starts to become challenged when we add dozens, hundreds, or even thousands of different part designs that move through these same machines. And that is exactly what we did as the twentieth century progressed. We grew our product lines to dozens, hundreds and even thousands of product offerings. This made sense from a business growth perspective; we needed more stuff to sell. And as long as we maintained reasonably large batch sizes, the standard costs of having one product model in the factory versus one hundred models were still based upon the same theoretical standard costs and low enough to convince ourselves we would stay competitive.

But this is where the "aha" moment in mass manufacturing occurred! This is where the system started to break down and all of the so-called hidden wastes of mass manufacturing really started to rear their ugly heads. Let's look at why all this diversity starts to erode the effectiveness of mass manufacturing. What does mass production look like to our example bracket in a highly diversified manufacturing environment? Well, this part is routed around the factory independently in spaghetti-like fashion to hit all of the machines and operations it needs to go to, just like every other part. And all goes fine as long as everything is working to plan in the factory and we have enough available machine capacity.

WHY WE BENEFIT FROM BATCHES IN MASS PRODUCTION.

Let's walk through an example of why bigger batches are more cost effective in a manufacturing system. A tenoner is a familiar machine in the woodworking industries; its purpose it to cut the ends of a piece of wood to shape and size. A double-end tenoner can cut and shape both opposing edges at the same time. You might use a tenoner to make the final sizing and edge shape on a kitchen cabinet door. It's a great machine, and there are some very flexible designs available today with shorter setup times, but in my experience the older machines could easily take one hour or more for an operator to setup and get right. Once successfully setup, they take just a few seconds to run each part.

So let's say in our mass manufacturing system we use fifteen seconds to run each part, and we decide to run two hundred forty parts through the machine in a typical batch. The setup in my example will take one hour. This was not an unlikely scenario for anyone who was in the furniture industry in say 1985.

This results in one hour of setup time and one hour to run the parts:

> 240 parts * 15 seconds/part * 1 minute/60 seconds = 60 minutes run time
> Set up Time = 60 minutes (fixed)

If our totally loaded cost for this machine is thirty dollars an hour for the machine depreciation plus labor—a reasonable estimate—we process these two hundred forty parts through the machine in two hours for a total cost of sixty dollars. This gives us a final cost of twenty-five cents per piece:

> (2 hours * $30.00/hour)/240 pieces = $0.25 cost per piece

Let's say that we decide to cut batch sizes so we can reduce inventories and build closer to the customer orders. If two hundred forty pieces could cover six months of demand, we could move to sixty pieces per batch and build only one and a half months of inventory, taking our inventory turns up from two to eight. This sounds like a great idea to raise our inventory turns by four times and help reduce our invested capital and warehousing costs. Now we are thinking leaner! But what happens with our cost with this change? Our run time for a batch of sixty parts will now be fifteen minutes:

> 60 parts * 15 seconds/part * 1 min/60 seconds = 15 minutes

Our setup is still fixed at one hour, so we use a total time of one hour and fifteen minutes (1.25 hours). And our cost is now thirty-seven dollars, fifty cents, for the total operation (1.25 hours * $30/hour). Our piece cost to run batches of sixty through the same machine is now thirty-seven dollars and fifty cents, for sixty parts, for a whopping $0.625 per piece:

> (1.25 hours * $30.00/hour)/60 pieces = $0.625 cost per piece

So with this "lean" change, we've more than doubled our cost from the two hundred forty-piece to sixty piece batches, $0.25 each compared to $0.625 each!

Further exacerbating this, a typical furniture component might go through ten to fifteen machine operations, just like this one. So it's easy to see that this smaller batch is going to cause a very serious cost increase as the part runs across all of the machines in its routing. This change in batch sizes will spiral costs upward unless we can find a way to take the setup costs down or find some other meaningful cost reductions that we can gain through the smaller batch sizes.

Does this sound like a bogus rookie example? Well, I've seen this actually happen in misdirected lean programs. To protect the innocent, the nameless shall remain so; however, whole teams of smart operations leaders have walked right off this plank, trying to rush to lean before the right preparations had been put in place. In one case the leader of the business stood firm by his efforts to cut batch sizes while his product costs went up drastically and, in less than two years, took the business into a severe nosedive.

Mass manufacturing is based upon the leverage that larger batches provide. Larger batch sizes allow us to amortize setup costs over more pieces, lowering overall costs. You might then ask, why not build even larger batches? Two primary reasons as you may remember for your studies of Economic Order Quantity:

1. The cost improvement is an inverse relationship. The cost improvement diminishes as we go up in quantity. Using the example above for even higher quantities, you can see the benefit tapers off as we increase quantities:

Batch size	Cost/Piece
60	$0.625
240	$0.25
480	$0.1875
960	$0.156
1920	$0.1375

THE LEAN HANGOVER

2. Larger batches create other costs and risks. Inventory-carrying costs are the most obvious. The cost of poor quality goes up if we create a bad part. We have larger parts and bins to store and move around. Floor space goes up. We can't process as many different parts over a given time. Our abnormalities causes even longer waiting times at machines because of the number of parts in each queue. So there is a point of diminishing returns.

> You might also be thinking, *Then why does lean manufacturing work to eliminate big batches if they provide lower costs?* That is because lean works to redesign the manufacturing system to eliminate these huge setup costs before you turn on smaller batch sizes. Think about this same analysis if we were to take the setup time of the tenoner down to a few seconds or none at all. We could then run very small batches cost effectively. Short setup times are one of the most important keys to lean manufacturing. And lean also produces the benefits of reducing other hidden costs that are created by big batches as discussed in more detail below.

But all does not go as planned. In fact, as complexity goes up, "working according to plan" can start to be the exception in the factory instead of the rule. Working according to plan generally acts inversely to the levels of complexity we put in our plants. In a diverse manufacturing environment, our simple bracket starts to hit speed bumps along its journey. What kind of speed bumps? Well, typical ones that might happen in a highly diversified factory. As we grew the number of parts on our shop floors, we ended up with more and more parts hitting the same machines at the same time, for example, and doing so in random order. As a result, we might exceed the machine capacity for a period of time. The result is that parts start to back up. We might have a machine breakdown. Parts start to back up again. We might find a quality defect in one of the parts on the assembly line and need to expedite parts through a set of machines to keep final assembly running. More parts start to back up. We might have an unusually high order that overruns machine capacity. Still more parts back up. These kinds of abnormalities and a host of others occur regularly in mass manufacturing plants. And, when abnormalities occur, parts queue up and wait at our machines.

Since we are having these random and unpredictable waits that occur across our machines, we look for ways to fix this. One of the more obvious potential fixes is that we can use overtime to catch up. Fine, but with a lot of complexity comes a lot of overtime. And because the abnormalities pop up at random places in the factory we end up with unpredictable overtime needs that move around the plant. Ultimately overtime becomes an ongoing ineffective and expensive use of labor. Overtime is inefficient and expensive and is only a good temporary solution. We need other solutions to take overtime down and run the business reliably on regular time mostly to meet our standard costs.

So what other solutions are there? Well, if we are having delays getting parts through the factory, where does that cause the greatest pain first? When parts aren't getting through the factory on time, we quickly shut down final assembly. We absolutely can't let final assembly go down. Shutting down final assembly means we don't ship finished goods, which means we don't generate revenue to run the business. This is not a good way to run a business.

So what's another logical solution? Well, we can build *safety stock* inventory to protect operations like final assembly. We might stock one or two batches of the components that we need to keep final assembly running. But wait a minute: don't we still have to run big batches to keep costs down? And don't we also have a complex product line with lots of parts? So, if we add "final assembly safety stock" inventory and possibly even add it across a significant portion of our product line, what does that do to our business? It requires big batches of a lot of part numbers. Now we've consciously decided to pollute the factory with batches of parts as safety stock, and we've also tied up more of our cash in inventory.

Whoops, not sure this is a perfect solution, but we have to keep final assembly running in mass production, so let's charge ahead! Great… guess we fixed that for now…

Another logical solution that we could use to avoid shutting down final assembly due to stock outs is to build in some of the wait times as normal steps in the part routings. That way parts can tolerate some abnormalities and still get to where they need to be on time. The theory of the case is if we build in some realistic wait times, we have a chance for part flows to tolerate slight abnormalities and get to assembly on time. The parts could even get to assembly early if everything goes smoothly in the factory.

So, let's think about this strategy for a minute. Aren't we still building with our big batches to keep costs down? And don't we have a complex product line with lots of parts? So what you are telling me is that we are going to build in these wait times. Doesn't that mean we are building in queues as part of the normal flow across our product offering across a ton of parts? So now we've decided to pollute the factory with more batches of parts as more work-in-process (WIP) inventory, and we've tied up even more cash in inventories because parts spend longer times on the shop floor.

These fixes all might sound logical, but in reality they start to produces a death spiral in mass production. Let's think for a moment about how long our cycle times might become with these two strategies. With all of these batches, safety stock, and wait times on the shop floor, it could start to take months for a part to make its journey from the starting process, through the factory, and finally into the customer's hands. Not only is this a poor utilization of money because our cash is tied up in hoards of inventory, but it also creates tremendous negative impacts through hidden quality issues, the inventory impedes our ability to make product improvements and engineering changes quickly, and it also produces huge potential obsolescence costs.

Get the picture? With strategies like safety stock and parts routing buffer times, coupled with the normal abnormalities that occur in manufacturing plants—machine breakdowns, capacity contentions, operator absenteeism, quality remakes, etc.—highly diverse industries that run with batch manufacturing inevitably ended up with pallets and bins of parts lining up behind machines, waiting their turns to get through each machine. In really complex factories we literally ended up with thousands and thousands of parts sitting on the shop floor, waiting to take their turns in the machining center queues. By the way, this is where the word "queue" came from as in "batch and queue."

Theses queues are a lot like the people waiting to get on rides at an amusement park... but not quite as much fun. *I'm starting to feel sick, somebody let me off this mass production roller coaster.* Or, as my fishing buddy used to say, "That dog don't run..." or, "Was that my hunting buddy?" But I don't hunt... I digress...

Batch and queue became the norm in so many mass production businesses because it was the only physical way to process parts and assemblies. And because the batches were large and we had so many batches across hundreds of diverse parts, the shop floor became completely clogged with work in process inventory. These shop-floor flows had become so complex that it was nearly impossible to control or predict all of the interacting flow conflicts. Basically, we had to hope parts and assemblies would get through the shop floor. The manufacturing "strategy" was to push parts through the system, plan for long lead times, and hope things got to assembly when we needed them to! This is exactly how batch and queue could cause the mass production system to spiral out of control.

Does this sound like bad science fiction to you? Well it's not. I've been in batch-and-queue furniture factories that had several hundred active models in their lineup and each of these furniture models had from, say, fifty to one hundred unique parts. Do the math, and you get huge number of components needed to manufacture all of these models; these factories could easily have fifteen thousand to twenty thousand discrete parts. Obviously not all of the different models of furniture were being built on the shop floor at any one time, but it was not unusual to have 20–40 percent in fabrication on any given day. Because of batch and queue manufacturing, there were so many pallets of furniture parts sitting on the floor in front of each machine that you could hardly walk the shop floor. Batch and queue production literally became a system of daily chaos!

With all of the inventory and the resulting long lead times in batch-and-queue manufacturing, our shop floor starts to feel a lot like the Pachinko games you find in a Japanese arcade. You know the game, the one where you put money into a machine to shoot metal balls up through a maze of obstacles, hoping they will fall into a magic hole in the middle of the game. If they do go into the magic hole, you receive even more metal balls so you can do it all over again. Receive enough metal balls, and you can trade them in for valuable prizes like a soothing pack of menthol cigarettes or equivalent. Except most of the metal balls never make it into the magic gate. So you feed the game your money, shoot the balls to nowhere, and go home with a thinner wallet or maybe some cigarettes. This is a very popular game in Japan. You can see why…

Batch-and-queue manufacturing often felt just like this. The parts in the factory were like the balls in Pachinko, trickling down through the factory, and assembly was like the magic gate hoping the right parts would trickle in at the right times! And in batch and queue, the results were about the same as in Pachinko—a thinner wallet!

This was the reality of running mass production, and it happened every day. As complexity went up in our factories, lead times to process parts also went up and became less predictable. Delinquents and schedule changes and expedites became the regular daily work in this chaos. In fact, in many of these factories, the only way parts could get to final assembly was for people to go find them and force-feed expedite them through the machines to get them processed. Factories regularly ended up with whole departments of people called expeditors just for this reason. Which, by the way, only added to the batch and queue delays by making other parts wait even longer for expedites to be processed. Not a fun way to run manufacturing. Diversity, complexity and large batches could topple the scales of balance in mass production.

MRP SOLUTION?

Sure there were systems to help handle the flow of parts through mass production factories—paper systems early on and computerized information management systems later. Systems that were supposed to rein this in and help make things flow on time.

Materials Requirements Planning (MRP) computer systems were supposed to make everything flow by controlling where everything needed to be based on machine capacities, part routings, and data that was regularly entered into the MRP system. This was a great idea and theoretically should have worked. Theoretically! The problem with MRP systems on the shop floor is that random machine and part disruptions occur all the time. Disruptions occur because a machine breaks down unexpectedly or an operator doesn't show up for work or someone doesn't put the right inventory quantity into the system when he finished an operation or an expediter pushes parts through the system because of an urgent need. These are not things an MRP system handles well; the programs are based on predictable machine capacities and part processing times and planning for those. These disruptions cascade down through the machine's queue, backing up parts across long streams of flow even in MRP run shop floors.

These types of problems and others were chronic in MRP systems. The millions of dollars we invested in MRP systems were not capable of handling all of these abnormalities. In the end most MRP systems just added a bunch of costs to the shop floor to ultimately be overridden by expediters with manual processes.

Not to continue to kick a dead dog, but there could also be significant quality problems hidden in all of this mass production inventory. With more and more parts moving through the machine processes, it can become nearly impossible to detect a quality problem when it is created. When a defect is created in a batch-and-queue system, all of the inventory ahead of and behind the problem can cause the issue to be "hidden" for long periods of time. A machine operation could create a defect that may not be detected until weeks or months later because it was buried in piles of inventory only to show up in later manufacturing steps. The batches and queues create a terrible lack of accountability. Who made the mistake, how and when, and what do we do about it now—two or three weeks later? There isn't a line-of-sight urgency to identify and fix problems that are buried in piles of inventory at any given machine. With all of the big batches and long lead times, quality problems were hidden, and detection was far removed from the creation of the problem. It doesn't take a rocket scientist to understand that the batch-and-queue shop becomes very tough to manage as we proliferate and add high levels of complexity.

Okay, you get the idea by now, and enough beating up batch and queue mass production. In its day, mass production was the lowest-cost way to manufacture products. Mass production was part and parcel with the industrial revolution and the advancement of the civilized world, raising our standards of living to levels never before seen. But in a way, mass production was actually becoming its own undoing as we drove more and more complexity into our factories. With more and more diversity and complexity on the shop floor, we had opened up the flood gates, and mass production was starting to make less and less sense as the most efficient way to produce. There was an opportunity for change and improvement. Enter Toyota and Taichi Ohno and lean manufacturing.

THE SEVEN HIDDEN WASTES OF MASS PRODUCTION

1. *Overproduction*

    Overproduction results from building something when it is not required. Large batch sizes and building to forecast are the main causes of overproduction. We call this "just-in-case manufacturing" instead of just-in-time manufacturing. Overproduction is imbedded in mass manufacturing systems because of its design around long setup times and the resultant big batches to be cost effective. Long setups mean we have to produce weeks or months of inventory, and that in turn makes it more difficult to match actual demand with production forecasts. The wastes of overproduction include hidden costs of poor quality, excess inventory costs, floor space costs, excess and obsolete costs, and long lead times.

2. *Waiting*

    Waiting occurs when parts are waiting in machine queues instead of being turned into products that we can sell to make money. In mass production systems, wait times typically make up more than 95 percent of the time a product spends on the shop floor. Wait time means we are not using our cash fast enough to get the customer's cash. Lean manufacturing links process together so that products flow, thus reducing wait times.

3. *Transporting*

   Transporting parts between processes is an indirect cost that adds no value to the product. Traditional machining centers are setup functionally so that product is designed to move from machine to machine. We add people to move these parts around, resulting in cost that adds no product value. The more we move products, the more they also get damaged. In lean manufacturing we link processes so that the movement is direct from one machine to the next.

4. *Inappropriate Processing*

   Remember the old cartoon with the giant machine that was used to cut a huge log into itsy, bitsy little toothpicks, one at a time. This is inappropriate processing. Mass production is built around single purpose machines in machining centers, and often we are forced to use what we have, these million dollar machines, to do small tasks. In lean, we design right-sized machines that are appropriate for the job to be done. Break complex processes into simpler tools that are linked together through flow. Reuse old equipment for smaller, less precise jobs. Use less expensive machines designed with fixtures to drill, press, plane, punch, and do other simple operations. These will reduce your capital investment and give you more flexibility to link small machines together to create flow.

5. *Unnecessary Inventory*

Large work in process and finished goods inventory are the direct result of long machine setups, wait times, and machine abnormalities, and they result in long lead times and overproduction. Wait times increase lead times, which increases batch sizes. We need to have enough inventory to keep up with orders until the next time we build this product. It's a viscous cycle. Excess inventory creates inventory carrying costs, hidden poor quality costs, and excess and obsolete costs. Work in progress clogs our shop floors and causes ineffective use of space.

6. *Excess motion.*

Excess motion is too much bending, stretching, walking, lifting, and reaching. This is all about ergonomics. Design your lean processes so the parts flow naturally with the operator's movements. Design part presentations so they are at the right height for the operator. Develop aids to help with lifting. Make part transfers direct and part of the operator's natural motion. This is common benefit of *Chuka-Chuka* cells. Use good lighting and visual presentations.

7. *Defects*

    We are all familiar with what defects are. Defects cause excess costs in rework, scrap, work stoppages, capacity loss, and a decrease in customer satisfaction. Work to eliminate them through fail-safeing processes or *poke-yokes*. Jigs, fixtures, and go/no-go gauges are ways to fail safe the process. Work to make quality defects rapidly visible by eliminating huge stockpiles of work-in-process (WIP) inventory.

ENTER "THE LEAN" SOLUTION

In *Capitalism, Socialism, and Democracy*, Joseph Schumpeter (1883–1950), an economist most recognized for his deep understanding of capitalism and its evolutionary development, defines capitalism as "the perpetual cycle of destroying the old and less efficient product or service and replacing it with new, more efficient ones."

It was after WWII that Toyota Motor Company and its employee, Taichi Ohno, envisioned a system that could make products more cost effectively than in mass production through faster flow in small batches. Demand levels in the post-war economy of Japan were low, and the economies of scale provided by big batches and mass production systems were not as effective as they once were in Japan, where the carrying costs of inventory are so high.

Ohno believed that many humble value streams, not a few mighty rivers, account for the great bulk of human needs. Ohno had studied and learned from Ford's Model T mass production assembly lines, but he recognized that under lower volume situations a different kind of production system was needed. Ohno concluded that his challenge was that he needed an efficient way to make a few products at a time, not hundreds or thousands, as was the case with Ford's Model T. He had studied the flow of goods in the supermarkets of Japan and believed that production should be more like that, where products were restocked in small batches to demand, not pushed out in huge quantities. And that work should be scheduled off of actual sales wherever possible.

Ohno may have been most visionary in his belief that consumer demand consisted of many small streams instead of giant rivers and that if he could design a cost-effective manufacturing system to build those smaller streams of products through smaller batches with shorter lead times, he could create a more effective manufacturing system. The previous chapter covered in detail how small batches are difficult to manufacture cost effectively in a traditional mass production system because of the long and expensive setup times. What Ohno envisioned instead was a method of continuous flow, where component parts moved through the factory in small batches and in rapid succession and directly into final assembly with very little machine setup and very little material handling. Since Ohno believed businesses would continue to sell more variety of products in lower volumes, he wanted to develop a way to cost effectively manufacture this same way; to build smaller daily or hourly batches that matched actual demand.

Ohno's overriding concept was to build only what was needed, closer to when the business actually needed it to meet sales demand. However, Ohno needed to devise ways to build these small quantities cost effectively to better match actual demand. What Ohno developed was a new system of manufacturing that was based on continuous-flow production in small lot quantities with little or no machine setup times.

To do this, Ohno needed to find ways to make continuous-flow, small batch production lines cost effective. At Ohno's suggestion, his colleague Shigeo Shingo went to work on setup and changeover reduction. Reducing setups to as close to zero time as possible could give Toyota and its parts makers the flow that Ford had achieved with his assembly lines but with the flexibility to cost effectively handle smaller batches, something Ford did not need to have with his limited product offering and huge volumes for the Model T. Ohno achieved this by designing quick changeover machines and smaller, zero changeover "right-sized" machines that could be placed in line to flow parts directly from machine to machine. The machines were designed to be simple to keep capital costs low while enabling processing steps to be conducted immediately adjacent to each other in continuous flow with little or no work-in-process (WIP) inventory. By creating very short setup time, continuous flow production, Ohno could finally build products in cost effective, small batches, and

...d to an entirely new level of high performance manufacturing, later named lean manufacturing.

Ohno's lean manufacturing system provided huge benefits beyond the traditional standard costs of manufacturing. When we manufacture with this direct flow, short lead-time, continuous production, we significantly reduce the seven hidden wastes of mass production that were covered in the previous chapter. Wait times are reduced by an order of magnitude because parts and product assemblies flow continuously through our factories. Quality defects that were hidden in huge mass-production batch inventories are uncovered almost immediately in continuous-flow production. Continuous flow also takes out all of the spaghetti-like transportation and indirect labor costs that we have with traditional machining centers in mass production. And, when we build products with continuous flow, we can build them in shorter lead times so we don't need to overproduce products to cover long-range sales forecasts.

Taiichi Ohno lived from 1912 to 1990. Ohno was first an employee of the Toyoda family's textile loom business, Toyoda Spinning, and began working at the Toyota Motor Company in 1943, eventually rising to the executive ranks there. Ohno developed these lean manufacturing methods after WWII and transformed these concepts into what became known as the Toyota Production System (TPS), now synonymous with lean manufacturing. Ohno published several books, the most popular of which was *The Toyota Production System: Beyond Large Scale Production*.

In what was considered a potential slight at Toyota, Ohno was asked to consult with Toyota's suppliers in his later years instead of progressing through the executive ranks. This may have been a result of his disclosure of the Toyota Production System to the rest of the world. He spent his later career teaching companies around the world about TPS through his association with the popular lean consulting firm, Shingijutsu, which literally means "New Engineering" in Japanese.

After the oil crisis of the early 1970s, Toyota cars quickly became popular for their great gas mileage. As Toyota started to grow globally, customers began to notice the unusually high quality and reliability of these vehicles compared to the vehicles from the big three automakers. Toyota's quality performance was so strong that they quickly surpassed the once behemoth industry dominant, General Motors, in customer satisfaction surveys. Toyota continued this trend, making strong sales and market share gains in the United States through the 1970s, 80s, and 90s. Toyota used their system of lean manufacturing to eventually takeover as the sales leader in the global automobile market. And with the rapid success of Toyota, business leaders took notice, wanting to know how they were producing such great automobiles. The manufacturing system behind these results, the Toyota Production System (TPS), began to catch the world's attention as the key to Toyota's better automobile manufacturing.

And that, in a nutshell, was the creation of lean manufacturing. Hopefully this history lesson gives you a high-level sense for what Toyota and Ohno were trying to achieve and how lean was conceived to be better system through building smaller batches, in continuous flow, closer to demand. In the next chapter we will develop a more robust definition of lean manufacturing so that we have a clear framework to understand exactly what we are trying to accomplish.

JIT AND THE JAPANESE

It is interesting to note that two primary driving forces that are unique to the country of Japan helped lean manufacturing to be developed there.

The first of these factors rarely gets the credit that it deserves in terms of what actually drove the lean manufacturing revolution. Japan is a country roughly the size of California, but with a population of 127 million people. California, in comparison, has a population of thirty-eight million people. And while Japan is about the same geographic size as California, it is estimated that only 20 percent of the land area in Japan is habitable because Japan is so mountainous. Thus, Japan is a very heavily populated country where space is at a high premium.

We've all heard about the extremely high costs of land in Japan, especially back during the bubble economy of the 1980s. This high cost of land, in general, results in an extremely high cost to store inventory. When I lived in Japan from 1992 to 1995, the businesses that I worked with ran their operations on the basis of as little inventory as possible. The entire infrastructure in Japan is built to address this space challenge with small trucks delivering small batches to grocery stores, retail shops, and manufacturing operations in short time intervals. It is just the way the Japanese have had to learn to compete in their densely populated country.

So when Taiichi Ohno and others in Japan were faced with the never-ending challenge of a lack of storage space, the Japanese had to figure ways to succeed within this environment. The solution was to find a way to cost effectively manufacture with less inventory, building products and components as close as possible to when the end customer actually needs them, and just-in-time manufacturing was naturally born out of this need.

The second unique factor that contributed to the lean revolution was the timing and depth of Japan's industrial rebuild after WWII. Japan's industries and cities were devastated by WWII, with many of them needing to completely rebuild. This caused tremendous industrial growth in Japan in the post war years. In America, by contrast, many of our large industrial investments occurred in the years between 1900 and 1945 before WWII. The Japanese were building their global prowess after 1945 and into the ensuing decades, and, as a result, benefited from all of the lessons learned of the great mass producers earlier in that century. They benefited by being a late bloomer through building on tried-and-true manufacturing methods, developing newer machine technologies, better computer systems, and better product designs.

When you put these two together, prohibitive inventory costs, along with the huge post WWII rebuild and re-invention of Japan, they resulted in the development of a better way to manufacture with much less work in process and finished goods inventory, or *lean manufacturing*. This was a

> capability that the rest of the world had yet to really find a need for. One can argue that the many side benefits of a lean manufacturing system and the reduction of the seven wastes in mass production were, to a large extent, by-products of Japan's basic need to radically reduce inventory levels and build to demand!

LEAN'S PRINCIPLES DEFINED

I remember at one point in my career sitting down with a very powerful CEO to explain why his business should drive a corporate-wide lean initiative. This was a business that had been a very successful batch and queue manufacturer for many decades but was also ripe with lean opportunities. The business had a few renegades who successfully started to introduce lean to their operations and portions of the leadership team were starting to catch on. My job as the corporate lean leader was to sell the CEO on a company-wide initiative. In my pitch, I walked the leader through some information on the philosophy of just-in-time manufacturing and building in smaller batches and all of the benefits that lean would bring to the business. However, I didn't have a really strong explanation on how smaller batch sizes and continuous flow manufacturing was actually accomplished. At the end of the session, being the excellent cost manager this CEO was, he was thoroughly convinced that lean was going to require a huge capital investment for all of these continuous flow lines, capital to buy much more dedicated equipment than what he needed today with batch processing, and that the costs of lean would drive him out of business. Ugh! I needed to better communicate how lean actually worked and how it could be cost effectively achieved.

In my early experiences implementing lean like these I was never able to put my hands on a clear definition of lean manufacturing. As I was living through the lean transition, a clear definition wasn't readily available through our consultants or in some sort of guidebook or short presentation. Lean was so much easier to teach through examples that most consultants preferred to teach by doing; you had to live through the experience to gain the knowledge. This was a valid teaching philosophy, along the lines of the "teach a man to fish" proverb, and it also made for attractive billable hours!

Now that we understand mass production and its potential built-in challenges, let's transition over and develop a detailed definition of what exactly makes lean manufacturing "lean." A good definition of lean will help leaders get their heads around these core concepts and enable them to set off in the right directions. However, I've also found that it's difficult to give an exact definition of lean; it is better defined through a set of directional principles that capture the essence of what we are trying to accomplish. So I will break it down through a set of lean principles.

If we get to the core of what lean is trying to accomplish it is this:

> ## LEAN PRINCIPLES
>
> 1. We manufacture through continuous flow production, eliminating batch-and-queue manufacturing. Parts flow through the factory in very small batch sizes though continuous machine processes, directly into final assemblies.

This is a start; however, this definition is not actionable yet. We still don't have enough depth to know how to do lean from this simple statement. So let's drill down to the next level to understand more detail on how to achieve lean. Here are my lean principles with two additional important principles added:

> ## LEAN PRINCIPLES
>
> 1. We manufacture through continuous flow production, eliminating batch-and-queue manufacturing. Parts flow through the factory in very small batch sizes though continuous machine processes, directly into final assemblies.
> 2. We flow manufacture parts and assemblies, with little or no work-in-process inventory, to move products quickly through manufacturing with processing times that are as short as is practicable.

> 3. Wherever lead times can be made short enough, we build parts and assemblies to actual demand, not to inventory

Now we are getting somewhere; these lean principles are starting to feel more directional. However, these high level principles are still missing the details of how we actually go about making manufacturing processes lean. How do we cost effectively make the transition from big batches in the old mass production world to cost-effective flow processes with short changeover machines to create lean flow? And how do we accomplish these with lower overall costs?

We need to drill down one more level to understand these deeper how tos.

> ## LEAN PRINCIPLES
>
> 1. We manufacture through continuous flow production, eliminating batch-and-queue manufacturing. Parts flow through the factory in very small batch sizes though continuous machine processes, directly into final assemblies.
>
> 2. We flow manufacture parts and assemblies, with little or no work-in-process inventory, to move products quickly through manufacturing with processing times that are as short as is practicable.
>
> 3. Wherever lead times can be made short enough, we build parts and assemblies to actual demand, not to inventory

> ### Lean How Tos:
>
> - We identify common part and assembly flows or "value streams."
> - Within these value streams, we design long machine setups out; we design new machines/processes with little or no setup time so we can manufacture small batches cost effectively.
> - We link these machines together in new continuous flow value streams that replace our old disconnected value streams with cost-effective, small batch, continuous flow production.
> - We move parts immediately from machine to machine with little or no work-in-process inventory so that products move rapidly through manufacturing, with short processing lead times.
> - These short lead times enable us to manufacture parts and assemblies to actual demand or closer to actual demand, not to long range production forecasts.

These are the fundamental principles of lean manufacturing, my key concepts to creating lean shop floor flow. Think of these principles in the following simple way: your purpose is to identify common part and assembly flows in your factory, reduce machine set up times, put the machines together and link them to create direct flow processes, enabling your manufacturing system to make products rapidly in small batches, closer to customer demand. That is a mouthful but it is lean in a nutshell.

Now that you understand what makes the lean manufacturing system work and how this is accomplished at a high level, we can start to develop the details of how to implement lean to produce better results in your business. However, before we jump into my lean lessons, let's cover one last piece of background on the lean manufacturing system, the specific benefits we gain out of this newer manufacturing method.

WHY LEAN IS BETTER

Following are the benefits that you will gain from a lean manufacturing system. These performance improvements should motivate you to commit and stick to lean manufacturing in your business.

DIRECT LEAN BENEFITS

- Shorter lead times.
- Improved inventory turns.
- Reduced floor space.
- Lower labor costs.
- Improved quality.
- Higher profitability and return on capital employed

If those aren't enough to have you salivating, following is a list of indirect or second-tier benefits generated by a lean system:

INDIRECT LEAN BENEFITS

- Improved worker engagement and accountability.
- Shorter recovery times.
- Faster implementation of product revisions.
- Improved new product designs.
- Lower scrap costs.
- Lower excess and obsolete costs

The direct benefits above are "direct" because they are areas where you will be able to measure progress directly as a result of your lean transformation. These metrics will improve "directly" with your lean progress.

The indirect benefits will show up in the system over a longer period of time, as the culture of your company evolves with lean manufacturing. These are not clear leading indicators of improvement but show up over time as the effects of lean manufacturing start to congeal.

This is a subtle distinction; one of these groups is not more important than the other. In fact, the secondary benefits may even be more important to the long-term success of the business. But the distinction does help to give us an indication on where to look to see that our lean program is working.

DIRECT BENEFITS

Shorter lead times: When we convert our plants to processes that flow continuously from machine to machine with little or no WIP inventory, the direct benefit we gain from lean is greatly reduced *cycle time.* The time it takes to manufacture a product end-to-end through the factory is called "cycle time" or "lead time." In lean we are going to design ways to manufacture parts and assemblies through continuous-flow processes with little setup time. Many machine setups will be completely eliminated, and others will be reduced to a fraction of what they originally were. As a result, lean will reduce process cycle times dramatically; in many cases cycle times are reduced by 80 to 90 percent. Building products businesses I've worked with have taken their manufacturing cycle times down from eight to twelve weeks to five days or less.

Improved inventory turns: As cycle times are reduced, inventory turns go up in direct relation. If it takes us three days after lean to make what used to take thirty, we have approximately one-tenth the inventory in the manufacturing system. We actually should have less WIP inventory proportionately because in lean we work to have as little inventory as possible between operations through direct, one-piece flow machine transfers. Work-in-progress inventory is the direct enemy of continuous flow manufacturing; you will drastically reduce work-in-process inventories if you are doing lean right.

Secondly, since you are making products with shorter cycle times, closer to when they are needed after lean, you will also be able to reduce finished goods inventory. If we can make a product in two weeks versus ten now, we need much less FG inventory to cover demand over those two weeks. If we can ultimately make all our products in a day or two, in theory we can start to build products to order and won't need finished-goods inventory at all.

As your cycle times come down, your inventory turns will go up. Your inventory turns will move up drastically if you get lean right. Batch businesses often have inventory turns on the order two to six per year. Through lean, businesses improve their inventory turns to the low to mid teens, ten to fourteen times per year is typical. CFOs like higher inventory turns a lot as do CEOs. You will be a hero, and the business will shower you with money and accolades if you can get this done!

Reduced floor space: We free up tremendous floor space through lean. Fewer inventories are lying around, taking up our precious space. Work-in-progress and finished goods inventories come down through short, continuous production flow. It's not unusual to free up whole factories or warehouses directly through lean.

You will also be more efficient with your machine space and cellular machine layouts. With your new continuous-flow lines, parts will move directly from one processing step to the next. You will also right size your machines; the lean philosophy is that machine size should be about the same as the part size. These will result in more space freed up due to effective machine design and layout!

Lower labor costs: You are going to setup processes to flow from machine to machine in manufacturing cells. Indirect labor to move big batches of parts around from machine to machine will be greatly reduced or eliminated. You will also waste less time on long machine setups, providing additional indirect labor savings. Labor costs come down as a result.

Improved Quality: Quality will improve because you don't have huge queues of work-in-process inventory everywhere. Lean's shorter cycle times and WIP elimination mean that parts go quickly into final assemblies where we do our final quality checks, so problems show up in minutes or hours instead of months. We call this *line-of-sight manufacturing*; an individual can see what they made and know if they did it right because it works properly in the final product within their line of sight. Workers can talk to each other when they find an issue to identify and solve problems quickly. Hundreds or thousands of parts aren't made before the problem is detected. Problems are not hidden in piles of inventory waiting to be discovered months later. The cost of poor quality decreases significantly as a result. Labor costs also improve as a result of not wasting effort on quality issues and rework.

Good lean process design also implements go/no go quality checks. These are called poke-yoke's, or "fail-safeing." Poke-yokes can be used in mass manufacturing systems too, but they are more prevalent and integral to a lean system. Think of it this way: since we design custom, right-sized machines, they are usually dedicated to a single or small number of operations in lean. It is easier to build in jigs and fixtures that fail-safe the dimensions as the operation is done. If you have a machine that is dedicated to one operation, you can build part nesting and quality gauging right into the machine to ensure the operation is done right. Integrated poke-yokes are an important and intuitive part of dedicated continuous-flow lines. We become better at implementing poke-yokes in lean and our cost of poor quality comes down even more.

Improved Profitability: Your profitability will improve as a result of all these combined performance improvements. Indirect labor cost reductions and quality improvements will have direct, positive impacts on your P&L. You will also see substantial additional cost savings generated by the indirect benefits covered immediately below.

Your return on capital invested will improve. Inventory turns go up, and money previously tied up in inventory comes out of the system. Labor costs and capital tied up in inventory both reduce, so return on investment improves. The level of improvement varies case by case; however, lean businesses can typically improve their return on capital invested by as much as ten to twelve points. That will certainly make shareholders happy!

Further, you can use this inventory reduction to pay for the capital investment you will need to make your right-sized machines and continuous-flow lines. In your lean conversion, you will be able trade inventory dollars for capital investments that generate all these benefits. It's pretty easy to see how this works; if we double our inventory turns from, say, four to eight in a one-billion-dollar business, we take one hundred twenty-five million dollars out of inventory. That is some pretty good change to use for capital improvements. Spend some of it on lean capital improvements, and put some of it in the bank!

INDIRECT BENEFITS

Improved worker engagement: As a result of the line of sight manufacturing that lean creates, worker engagement and accountability improves tremendously. People want to contribute and do the right thing at work when given the opportunity to do so. But, in a mass production system problems are hidden and remote from the source because they are buried in piles of inventory. It's hard to hold people accountable for a problem that was generated months ago, and attempting to do so can lead to a lot of finger pointing because the potential root causes are so disconnected from discovery and may even have disappeared by the time of the discovery.

In lean there is no place to hide, and that is a good thing. Since we have little inventory, problems surface quickly, and we also correct them quickly. This is good for the business and for its workers. People want to do the right thing, and when we give them a strong chance to do so, they will do even more of the right thing! Lean gives them just that—a higher probability of success. As a result, workers are happier, management is happier, and engagement improves. We all make better products. This is a win-win!

Shorter recovery times: Recovery times are much smoother in a lean manufacturing system. Businesses with highly configurable product offerings in particular can significantly benefit from shorter recovery times. Kitchen cabinets, windows, and doors are a couple of building products that fit this category. These products can and do get damaged at the building site; it happens with all of the materials moving around during construction activities. Or, in another typical scenario, a dimension gets taken incorrectly, and the product needs to be reordered. In these cases the customer needs a quick replacement; they can't wait eight or ten weeks for a new product. In a mass production system, we needed to expedite replacement orders to hope to get it through in short lead times. Mass production systems end up with lots of these kind of expedites to the point where they upset the flow of normal production. In a short-cycle time, lean system, we can build the replacements alongside our regular production orders. Our lean lead times are short enough to do this. Recovery times are much more seamless in lean businesses.

Faster implementation of product revisions: Product changes can be implemented faster and at lower costs in a lean system. Let's say we identify a potential performance improvement opportunity with a product in the field, and this particular opportunity requires that we make a design change to the product. In a mass production system, we have a substantial amount of both finished goods and work-in-process inventories. What do we do with all of this inventory? Do we rework it, scrap it, or use it? These can all be costly, long-cycle time options.

In a lean system we have little inventory. We make the changes, use or dispose of the little inventory we have on hand, and are back up and running. Reaction times and costs are greatly improved. This is a big improvement for businesses with riskier and frequent new product introductions.

In the same way, lean's short cycle times also enable your business to make more frequent product introductions and enhancements. Your productivity changes can be implemented much faster. Successful businesses depend on driving productivity initiatives every year to take labor and material costs out of their products. These productivity improvements often require engineering changes to a product or process. In a lean system, you can make these changes much quicker, often on the fly, because you don't have the inventory to work through like you would in a mass production system.

Better new product designs: Shorter production lead times and little work-in-process inventory will help you become a more agile new product designer. Your product designers are not hand cuffed by mountains of inventory that get in the way of implementing good design ideas immediately.

You will also become really great at designing products that go hand in hand with lean processes. Product designers begin to design parts that fit a lean manufacturing flow because that is how the business operates now. This is a how we grow and sustain our lean processes. When we teach our design teams what matters in lean; they are held accountable to finding ways to make their products in short lead times. This lean capability leads to better product designs because they have lower costs and shorter lead times designed into them.

Lower scrap costs: With lean, you are going to have fewer quality problems overall because of the line of sight manufacturing and accountability. Also, with lower inventories and improved reaction times, you are going to have a lot less inventory to scrap out if a quality problem arises or a design change occurs. Because of these two reasons, lean yields lower scrap costs.

Lower excess and obsolete costs: Sometimes you just can't sell everything that you made. Customers didn't like the product, the competition had a better one, something was off with the functionality of the product, or you decided to replace it with a newer product. Similar to scrap costs, in a lean system, your excess and obsolete costs will be lower. You have fewer inventories to dispose of when these bumps in the business road occur.

That is a long list of benefits. No wonder manufacturers have made lean a top priority!

There is, however, one additional significant benefit that was left off my list for last. It was left for last because it is the overall benefit you get from the combination of all of these improvements. The combination of all of these benefits creates a *lean growth engine* for businesses; your improved performance dramatically increases customer satisfaction, and that, in turn, will generate sales growth!

Let's look at this from the customer's perspective:

1. Customers benefit from shorter lead times. They want their products when they order them. When you outpace the competition with shorter lead times, you will be stronger at order fulfillment, with fewer stock outs to put yourself ahead of your competition.

2. Shorter lead times mean that when something goes wrong, your reaction times are shorter. You earn loyal customers through rapid, effective recovery when something fails in the customer's eyes.

3. Over time, your price/value relationship will become more competitive because your costs improve under lean. This is good for your profits and good for your customer through better value propositions.

The end result is that a lean enterprise is much better at serving its customers. Satisfied customers come back for more business, and new customers come to the business because of its great brand. Using lean to improve your business performance generates strong growth. And once you are ahead in this cycle where lean performance fuels growth, it's nearly impossible for your competition to catch up! This cycle of lean business improvement driving customer satisfaction and sales growth has been referred to as the "*Lean Growth Engine.*" The business grows directly because of the improvements made through lean.

So now that we know how lean manufacturing improves on mass production, it's time to switch over to the core of my book, my thirteen lessons on how to get lean manufacturing right. These lessons contain what I have found to be the keys to unlocking the full potential of lean; the core approaches that lean leaders use to succeed.

> ## HOW TO MEASURE LEAN IMPROVEMENT?
>
> If you had to monitor a few metrics within your financial performance to see if lean is working, what would they be? I've been challenged with this question a few times, and it is a tough one. Senior management needs to be able to connect cause and effect and show it to their board and the shareholders, so they can continue to have support to make lean investments. You can't very well ask for a leap of faith as you make the lean change, just on the backing of others that have made it work. We look for the few salient signals we can count on.
>
> The thing that changes most directly through lean is your cycle times. But cycle time is not the cleanest of metrics. It can be difficult to accurately measure; it varies greatly from product to product, and it's not something shareholders might know how to value. Lead times might be the best cause and effect lean metric, but lead time needs to be part of a set of metrics that can be used to track lean progress.
>
> The following four metrics are the ones you should monitor to track lean improvement:
>
> 1. *Cycle Times*
> 2. *Inventory turns* will improve up over time. This metric is going to move slowly, but after twelve or eighteen

months, it should show a continuing upward trend if you are being aggressive with lean implementations

3. *Productivity.* We take labor and material costs out with the shop floor improvements. Actual cost per unit must come down.

4. *Profits* will improve. However, in many businesses profitability will not improve overnight. You have to balance the pace of your capital investments with a depreciation schedule that your P&L can handle. It is going to take several months and maybe even a year or more of developing and implementing flow lines to show significant improvement in profitability.

No one or two of these can capture the improvements alone. That is why it is important to watch all of them.

PART 2

ENABLING THE LEAN SYSTEM TO WORK

LESSON 1:

THE LEAN FAB FOUR

> The secret to lean manufacturing is investing in key systems and resources who convert the business to lean and keep it there.

Here it is, folks. Here comes my most important key to getting lean manufacturing off the ground and headed in the right direction, and I'm putting it out in the open, right here, up front and in my first lean lesson. All here in print for absolutely nothing extra (since you've already paid for the book…). No extra fees, no small print, no hidden costs. Wow, what a bargain! This way you don't have to wait until deep into the book to see where I am going with this lean thing! I figure you can decide right now if you want to sign up for the snake oil I am selling or hawk this book on eBay and go read something interesting by Jim Collins instead.

You've heard of lean manufacturing referred to as a system. Well, lean is absolutely that—a system of manufacturing processes in its end state. However, the secret to successful lean isn't simply understanding how lean manufacturing behaves on the shop floor and then mimicking that; it is knowing how to build an organizational system that can create cost effective, reliable lean manufacturing flows. What enables manufacturing to get the point of really being lean is the leadership commitment to creating a system of planning product designs and their respective production processes to be lean 100 percent of the time! This is the critical system behind the lean manufacturing system, and the one that businesses botch most when they struggle with lean.

The key to lean manufacturing is the system of designing and planning production processes to be lean so that they actually are lean when you do the manufacturing. That seems obvious doesn't it? Almost like one of those "duh-dads" that I still get from my youngest daughter when I say something that is way too obvious. Apparent as this seems, it is surprising how many leaders still botch lean by starving the system of resources that actually design, implement, and support their lean manufacturing processes. You can't do that and have successful lean manufacturing. You will fall short every time, I *gaar-un-tee it*, as the saying goes.

Okay, then the question is what is the makeup of this organization that can proficiently design lean manufacturing systems, and how do we create those capabilities? This is the topic I take on here in my first lean lesson.

There is a set of key organizational capabilities that are both a requirement to and starting point for consistent lean manufacturing. There are four parts of the organization, my Lean Fab Four, which must all be capable and rowing in the boat together to make lean work well. The following are the four functional disciplines that are the true enablers of successful lean:

1. *Product Engineers* that don't lock big batch processes into their product designs.
2. *Process Engineers* that can design and implement continuous flow production processes.
3. *Information Technology* processes and data flow that enables, supports, and enhances continuous flow processes.
4. *Material Management* processes that can efficiently and reliably deliver parts to the lean processes that this team designs.

Lean manufacturing is an unrelenting, never yielding commitment to only put processes on the shop floor that are engineered to flow fast and nearly flawlessly. That is exactly what it takes to be successful with lean. So, how do you do that, Mr. Rocket Scientist? Well, you have to commit to and develop resources that do this work; yes, I mean human resources—those animated little things moving around your business we call "people." And, I'm sorry to poop on your near-term profit parade, but with the state your current manufacturing may be in, you may have to commit more additional resources than you might have the stomach for, to re-invent your business and make it competitive in a lean world!

For successful lean manufacturing, businesses almost always need to invest in new human resources or re-allocate existing ones to the four functional areas listed above. I don't care whether you choose to invest newly or reallocate, you can pick your poison, but you will need to have strong capabilities in each of these functional areas. Obviously reallocating resources is the easier and less costly of the two as long as you end up with the right talents in the right places. The point is that you are going to have to deliberately transform your organization to create strength in these four disciplines and support them to get lean right.

Yes, believe it or not, I am advocating that leaders actually lead by taking some risks and investing in their capabilities to design a better manufacturing system. Mass production systems were lethargic, slow moving behemoths; they didn't need the same intensity of focused resources to do this work of making continuous flow manufacturing systems which function nearly flawlessly all of the time. The idea with lean is that you invest in select areas of the business to generate a payback through more efficient operations, increased customer satisfaction, and top-line growth. Novel idea for you blockheads that think the definition of lean is broad cuts in your professional staff.

Let me cover that fork in the lean road here. My definition of lean in this book does not equal cutting your professional staff to become "lean." Many businesses accomplish their version of lean by cutting their overhead staff and then pushing the remaining people to do more, including trying to make their shop floor lean. With this system, everyone in the business walks around professing "how *lean* the business really is" and, remarkably, how they can't get anything rationale done because the business has so few human resources. So everyone runs from fire to fire, waving their arms in the air, and screaming for help. Refer to "private equity investors" for more on this "lean" approach.

That is a different kind of lean, and if you want to run that playbook, fine, but you are doing that for a different reason than trying to be successful with lean manufacturing for the long term. What you get with this technique is a lean overhead cost structure and usually lousy lean manufacturing. Why? Because there is no one to do the work of creating and sustaining lean product and process flows. This is a short-term strategy—penny wise and dollar short.

Yes, for all you manufacturing leaders collecting the big paychecks, those of you who got to the top of the heap because you are good at being stingy and getting more for less from your people, you need to decide if you really want to commit to this lean manufacturing concept. If so, then you need crack open up the company wallet (it may be stuck closed because the leather sides have hardened due to inactivity), turn it over gently so it points down, and spill a few bucks out of it. You need to spend a few of those clams to resource the four areas that actually do the work of planning and preparing to be lean; novel concept, Einstein… actually investing in people to get something! Maybe they'll even make you CEO once you show the big muck-a-mucks you can take on a *big* risk like this and actually succeed at something… finally…

Let me briefly discuss the roles each of my fab four functions plays in lean manufacturing and what is different about them in lean:

1. *Product engineering.* Product engineering is the business of designing new products, improving existing products, and driving productivity projects. Product engineers must enable lean process flows by designing products that are not locked into huge, monument, long-lead time, batch processes. Lean product engineers will work with your process engineers to evaluate and choose early designs concepts that will flow quickly and reliably and not be bound by long changeovers and unreliable processing that causes huge inventory stockpiles. Lean product engineers must commit to designing products that move through the factories in short lead times with continuous flow.

    My Lesson 7, *Lean New Product Development* and Lesson 11, *Three Product Design Strategies Make Toyota a Great Lean Manufacturer*, provide in-depth discussions of how product engineering successfully supports lean manufacturing. Product engineering is critical to getting lean right, and it is this collaborative product-process engineering relationship that creates effective integrated lean flows on the shop floor for every product design.

2. *Process engineering.* Process engineering is the most important of these four functions within the lean manufacturing system. They are the hub of lean manufacturing process design activity. At the core of highly effective lean businesses are process engineers

and machine experts that design and specify low-cost, small-batch machines, which are linked together through effective part transfers to create highly capable, continuous-flow processing lines. Complex manufacturing businesses need the capability to create smooth, continuous flows out of historically chaotic part flows. Great lean process designers are able to envision and implement effective methods of lean flow, creating unique and innovative solutions that can't be bought from a catalog or off of the Internet. Highly successful lean manufacturers are able to design these "custom-flow" manufacturing systems themselves and implement them on the shop floor.

Your process engineers do this work. They become the heart of your lean manufacturing system, designing part flows and assembly flows that move quickly and reliably through the plant, just like blood flows through our veins.

Process engineering also quarterbacks and coordinates the work of all four of these functional areas to create lean manufacturing processes. They work with the product engineers to translate product designs into lean processes. Likewise, they work with the IT group to ensure that the necessary data flows are designed and ready to support the shop floor product flows, data from the order management system that is translated into pieces and parts requirements and data for production control to manage how parts and assemblies move through the shop floor. Lastly, they coordinate the inventory management systems and material flows with your material management people. Process engineering is at the center of designing lean manufacturing systems, managing the overall project, timing, and implementation of lean shop floor projects.

My Lessons 2–4 cover the detail of process engineering's role in lean manufacturing and provide an organizational model called *Manufacturing Systems Engineering* for managing process engineering to create successful lean.

3. *Information technology*. If process engineering is like the heart in your manufacturing system, and your part flows are like the blood flowing through the factory's veins, then your information systems are like the central nervous system that guides all of the actions that take place on the shop floor. Information and data flow affect a manufacturing system from top to bottom. It starts with the bill of material (BOM), extends to the order entry system, to the inventory management system, to production control and to the shop floor data systems, just to name a few. Your product and process engineers will work with your information technology team to design the information flow requirements of lean manufacturing processes, and then your IT people will need to be capable to execute to these specifications. Information technology must enable lean flow, not get in the way of it. Poor information systems will cripple your lean manufacturing efforts. And all of your information systems, from top to bottom have to work nearly seamlessly and consistently to create reliable, short-lead-time production. If

they hiccup and sputter, so will your lean production system.

The decision to convert to lean has to be enabled through capable IT systems either via existing systems or by upgrading to new systems. We would all love to have fully supportive IT systems when we embark on the lean transformation, but that is rarely the case. Mass production had such different shop floor data requirements that it's unlikely those systems will be ready for the lean change. The lean conversion usually forces the need to upgrade information systems and to do so early on so that we can move forward with the lean transformation process. Systems need to support lean from the start and evolve with it in lock-step.

Frankly, this is an area where businesses often struggle to make adequate investments and do so early enough alongside their lean conversion. The reasons behind this are that new data systems are usually expensive, they require huge levels of resources to convert from old to new systems, they do not have an easily quantifiable payback, and systems conversions are fraught with risk.

Information technology systems are an infrastructure investment, one where it is difficult to quantify a direct payback. Part of the debate is determining how much savings a new IT system is going to generate so we can scale the investment accordingly. Information technology systems can be very expensive, especially the big ERP systems who shall remain nameless here. Obviously we want spend the smallest amount we can to get the functionality we need. In fact, these big ERP systems cost so much it's hard to see how anyone can get the bang for the buck, but I will leave that analysis for you to do. So, the question is, do we buy all of the functionality of the big ERP system or go with the bare-bones in house developed system or somewhere in between? It's hard to connect the added costs to added benefits but also risky to buy down too far.

And since the IT spend is going on at the same time that the manufacturing system is being upgraded to lean, we get a double whammy. The price tag for both of these projects can really test the CAPEX budget. With big-ticket items like this, investors and boards can and should be wary—how do we know how much lean is worth to us; are we sure we are going to get the payback for all this lean investment? This part of the lean journey can become a bit of a leap of faith. Have you ever met a top business leader who enjoyed walking to the end of a plank and waiting there for twelve to twenty-four months for the results to start to show? Lean can do that to you...

The process of migrating to new systems is also very risky. The business has to keep running as usual without missing a heartbeat while it moves from its old to new systems. Getting information systems changeovers right takes time and costs a lot; getting them wrong costs even more. Unfortunately, I've helped clean up some of these mess ups, and they are not fun. And sometimes poorly executed systems implementations can't be cleaned up fast enough, resulting in serious financial damage to the businesses. The point is that this need for improved data systems and data flow often adds another challenge and risk to the lean conversion. These require serious financial and resource commitments. Impatient CEOs and investors rarely have the stomach for these high investment costs, especially those looking for short-term returns or are running businesses already in financial stress. Investments for new information systems coupled with lean manufacturing capital investments can make the lean transformation process a tough sell to your investors, so tough that it can cause business leaders to shy away from making these needed information technology investments.

The point is, don't be a stone head and saddle your lean conversion with the burden of old, inadequate information systems. You can't achieve solid lean manufacturing while crutching along poor information systems. You need to address both of these needs or your lean efforts can go belly up. As I said in the Foreword, your systems are giving you exactly the results they are designed to give you. If you don't like your results, change your systems. Businesses that make the lean leap while hoping they can crutch their old information systems along get lousy results, sometimes even worse than their mass production performance levels.

My Lesson 6, *Beware the ERP monster and Lean Shop Floor Flows* provides additional discussion on several other information technology topics and their role in lean manufacturing.

4. *Material management.* In the lean organization, the recipe for materials management (raw materials and work in process parts flows) should be determined early on through collaborative product and process design activities. The product engineers, process engineers, and materials management people perform this design work during 2P and 3P kaizen events (see Lesson 7). Your process engineers will quarterback the design of efficient manufacturing flows, including the flow of raw and work in process materials. Lean material flows must be built in as a part of lean system. Well engineered material flows should shadow well engineered process flows.

In order to have reliable material flows, we need material planners that understand the blend of parts movements and information movements. This needs to be much more thoughtfully planned than the hope that I've seen used in many manufacturing systems. Material flows needs to be planned out to the last detail with carts, racking systems, the timings of flows and the sizes of manufacturing buckets. And the business needs to have indirect labor planned into the system to move materials around, often called *water spiders* in lean manufacturing. We can't use the assembly line supervisor or unspecified individuals to manage material flows; these need to be defined resources with defined work in the lean factory.

You will need depth and commitment in each these four areas to fully support lean. These four functional areas are like the four legs of a stool: if of any one of them isn't capable of supporting the lean load, the functionality of the entire system will likely be crippled! You will not be successful with your lean transformation without each of these functional areas resourced, empowered, and engaged in the lean changeover.

I've seen these short cut attempts to achieve lean manufacturing more than a couple of times and the results are not sustainable; leadership that tries to see if they can get the benefits of lean manufacturing in the short term and do so cheaply by working within their existing poor performing organization and systems. They direct a small team of resources to figure out how to make lean work for an isolated part of their operations. This is what I call the *build it one-brick-at-a-time* approach to lean, and here's my little secret: this approach never works for the long term! The business never builds the infrastructure to get completely over the lean learning curve and it never creates the business processes and systems necessary to sustain lean gains. This is a Trojan horse for serious lean and is usually done to appease upper management or a board or investors by leaders that aren't willing to take the risk or put the energy into reinventing the business for the long term. Here's my portrayal of how this one-brick-at-a-time lean conversion usually plays out. See if any of you can identify with this weak approach:

Let's carve away one of our good engineers, Engineer XYZ, and put him on line ABC because it's really a mess and see if Engineer XYZ can fix ABC through this lean manufacturing thing. Everyone is talking about lean like it's the greatest thing since PB&Js, so let's use Engineer XYZ to see what lean can do for us here at ZZZ Corporation

Tell everyone else in the business that they need to give Engineer XYZ all the support he needs, no-holds barred. This will be our lean learning line, and we want to get it right. Then we will have Engineer XYZ take what he learned on line ABC and implement it on all of our other lines including line AAA through BBQ, and don't forget line PDQ over in the Walla-Walla plant.

Engineer XYZ is a good soldier, and he works hard to improve line ABC. His activities involve improving the basic line layout, balancing and retraining labor, and making materials management improvements among others. Engineer XYZ finds out that most of his improvements are accomplished, however, through his manual labor and brute force. When parts are needed for the line to keep running, Engineer XYZ finds them. When a sub assembly breaks and has to be remade, engineer XYZ does the leg work to communicate the replacement parts need to the parts making area. When machines break down, Engineer XYZ grabs a tool box and fixes it. And this list goes on and on.

Engineer XYZ puts a lot of his own sweat equity into line ABC, working twelve to fourteen hours a day to breathe new life into it. He works with a team from his line to document major causes of line stoppages and flow imbalance, prioritizes those, and then works down the list to improve what he can to the best of his abilities. He looks for help from IT with systems improvements that could dramatically improve the performance of the line, but they are too busy to help Engineer XYZ with hundreds of work requests already in their backlog. Besides, line ABC needs big investments in information systems, and Engineer XYZ knows those will never gain funding. So Engineer XYZ moves down his list to work on manual systems and other activities that he can use to help the line flow in the short term.

Over time Engineer XYZ does successfully bring line ABC to higher performance levels through his short-term, brute force efforts. So, Engineer XYZ's boss schedules a big meeting with everyone attending, including the Grand Puba himself, to show how lean really can help the business. Yeah and rah-rah!

The auditorium explodes in wild applause after the cost improvements are presented, not for the manufacturing improvements but primarily because of the donut induced sugar rush and because it's Friday, and we all just got an hour off and are closer to the weekend! The organization has been down this path before and know our manufacturing still stinks, and everyone is ultimately glad this episode is over so they can get back to their day jobs.

As the next step, Engineer XYZ gets a big promotion and gains the official title of *Continuous Improvement Engineer* and is placed in the plant organization structure to now convert all of the manufacturing lines to this lean manufacturing system he has perfected.

In their infinite wisdom top management oversimplifies and decides this lean thing should be a piece of cake to get done in *all* the plants next year, based on Engineer XYZ's demonstrated quick results. Therefore the business had better build these lean performance levels into all of our business plans for next year! That way the board will be happy with our plan, and the plants will be held accountable to get lean implemented ASAP! And let's face it: careers are at stake here. If the op's leader can't figure out this lean thing in the near term, he's dead in the water. *Whew, glad we got this lean thing figured out…* is the collective thought bubble from the executive offices. Check that box off the to-do list.

Hoo-ray! We are all saved with this masterful plan!

F-me, would you please! Who is the brain surgeon that is running this place, anyhow? I wish I had a butt to kick in front of me right now. Have any of you experienced this charade for strategic manufacturing management? Unfortunately, I can say that I have.

So how does the lean process improvement strategy play out? Engineer XYZ plugs away on his next line and makes some progress there too. Goodie goodie gum drops, the change process is working.

But meanwhile line ABC back at headquarters is going exactly the opposite direction, reverting right back to where it was before Engineer XYZ paid special attention to it. As soon as the special attention goes away from line ABC, it slips right back to the performance levels it had before these focused lean efforts. Go figure.

Our systems are giving us exactly the results that they are designed to give us…

The lean conversion is affected by a large-scale commitment to this new and better way to engineer your manufacturing processes, and success requires change in your organization too. You will never reach critical mass with your product designs, process designs, information systems and material flows and make them work successfully and sustainably if you try to achieve lean by adding just one or two engineers at a time to magically create lean flows as isolated shop floor projects. You can't get there one brick at a time. It takes focused capabilities in all four of my fab four - capable product engineering, process engineering, information systems, and materials management - all in the boat and rowing together. Businesses that don't recognize these four pillars of the lean conversion rarely succeed. At best create they create lean flows that limp along. At worst, they can fail completely and often bring the business to its knees.

The secret to lean manufacturing is making a commitment to invest in your systems and resources so they can convert the business to lean and keep it there. This is the only way you can reach the goal line. The lean conversion goes go well beyond the fab four functional areas, but these four are the core. Without those four capable, empowered and headed in the same direction, you don't have a fighting chance of being really good at lean manufacturing.

LESSON 2:

LEAN IS CREATING CONTINUOUS FLOW OUT OF CHAOS

At the core, lean manufacturing is about developing cost-effective ways to continuously flow parts and products quickly and reliably through the factory.

The mechanics of the lean transformation have frequently been taught by consultants through rapid changeover kaizen events. Lean teachers use these rapid Kaizen changeovers to demonstrate lean's effectiveness to businesses that are starting down the lean path. The consultants swoop in and quickly identify simple, common part flows, the machines are rearranged through a consultant led kaizen event to create flow, and, voila, within a few days the business is out of the gates and well on its journey to becoming lean!

I've always been concerned with this lean teaching model. These quick kaizen fixes tend to focus only on our simple shop floor flows because this is where these techniques are effective. Lean consultants are expert at finding portions of the value chain where lean can be quickly demonstrated. However, these simple shop floor flows, where Lean manufacturing is virtually "plug-and-play," are usually only a small part a business's shop floor flows. These rapid improvements are viable teaching methods for simple flows or isolated parts of the value chain; however, the approach often lacks the engineering depth and long-term ownership needed to solve tougher shop floor challenges.

Most businesses are faced with parts of their manufacturing that are much more challenging to convert to lean's continuous flow processes. Businesses usually contain chaotic spaghetti-like part and assembly flows where we can't just simply link machines together to instantly create continuous flow. They may contain processes with huge setup times, or wait times for curing that slow down shop floor flow and keep us from being able to implement continuous flow production. There are areas like these on most shop floors that require longer range planning and product/process development to convert them to true continuous flow manufacturing.

Let's leave these quick lean changeovers for the minds that want to capitalize on selling simple solutions. If it were as easy as putting machines together in a row and punching the "On" button, we would have all figured out lean a long time ago; lean done, end of story, close the book! The reality is that implementing lean across the entire shop floor is more challenging than that.

Further, the more typical case, and the one that has not always been a quick success example for the lean consultants are businesses that have many thousands of parts and final product skus, with many of these having very complex and distinct part flows. Very few parts and assemblies follow enough of the same steps in succession such that we can simply put a set of machines together and create direct A, B, C, D flow to make lean instantly viable. To understand how to untangle these complex flows, we need to understand deeply how lean works and then take some time to study and make it work effectively for every product flow. We can't just flip a switch and suddenly change from building big batches to being capable of building small quantities cost effectively.

Some businesses and business leaders have made this mistake, thinking that all they had to do was sort of rearrange their machines and take the batch sizes down to create flow, and all of a sudden they had a lean manufacturing system. Businesses designed around big-batch processes can flounder if they don't pull the right lean levers at the right time. Lean leaders who jump into small batch processing too fast without developing the right machine and organizational capabilities can upset the delicate balance of cost and flow on the shop floor

As a result of not effectively dealing with these relationships, there have been many businesses that have struggled to effectively take their lean programs very far, businesses that have complex process flows but simply aren't able to develop the process solutions they need to create continuous flow across the entire shop floor. These businesses need a way to develop cost-effective direct product flows, yet these solutions are not as obvious as the direct A, B, C, D one-week kaizen pill. For many of these businesses their lean initiative sputters. Lean sputters because the business is fundamentally missing a deep capability to execute the lean principles I laid out earlier:

LEAN PRINCIPLES

1. We manufacture through continuous flow production, eliminating batch and queue manufacturing. Parts flow through the factory in very small batch sizes though continuous machine processes directly into final assemblies.

2. We flow manufacture parts and assemblies, with little or no work-in-process inventory, to move products quickly through manufacturing with processing times that are as short as is practicable.

3. Wherever lead times can be made short enough, we build parts and assemblies to actual demand, not to inventory

Lean How-Tos:

- We identify common part and assembly flows or "value streams."

- Within these value streams, we design long machine setups out; we design new machines/processes with little or no setup time so we can manufacture small batches cost effectively.

- We link these machines together in new continuous flow value streams that replace our old disconnected value streams with cost effective, small batch, continuous flow production.

- We move parts immediately from machine to machine with little or no work-in-process inventory so that

> products move rapidly through manufacturing with short processing lead times.
>
> - Theses short lead times enable us to manufacture parts and assemblies to actual demand or closer to actual demand, not long range production forecasts.

The key difference between lean and mass manufacturing is the ability to flow parts very quickly through the factory in small batches. My Lean How-Tos are the key manufacturing capabilities that you need to develop to enable your lean program to be successful across the entire business. Complex manufacturing businesses need the capability to create smooth, continuous flow out of historically chaotic part flows to convert to lean manufacturing. One of the greatest shortcomings, if not *the* greatest shortcoming, of weak lean programs is the inability of the business to create custom, cost-effective, short lead-time shop floor flows. I can't stress this enough. Making lean work in your business is going to come down to how well your organization can execute these Lean How-Tos!

So how do we get this lean transformation done? At the core of highly effective lean businesses are process and machine experts that do the work of creating exactly what my lean principles state, and they do it extremely well. The most successful lean businesses develop internal engineering expertise that can create low-cost, small-batch machines that they link together with effective part transfers to make highly capable, continuous-flow processing lines.

Highly successful lean manufacturers are able to design "custom-flow" manufacturing systems themselves and implement those on the shop floor. Study any successful lean model deeply, and you will find that they have learned to manage their own process designs through a strong machine understanding and knowledge base. Toyota was known as a great machine innovator with textile-weaving methods well before automobiles. They carried these process roots with them into the automobile business as a part of their culture. Lean leaders develop a *core-process design competency* that is expert at creating flow-manufacturing processes. This is the best way to ensure success with lean!

In my experience, the capability to develop custom lean manufacturing process solutions to smooth out and simplify complicated mass production flows is *the* most critical competency that effective lean enterprises must develop! This is one area where you have to make a strategic and deliberate commitment to invest to change your business!

Businesses love to design unique and innovative new products. These are the lifeblood of business growth and sustainability. But few truly are passionate about investing in "mundane" process engineering; process engineering is a necessary but lackluster capability in most businesses. However, consider this that same capability, the creativity of great product engineering, but now applied to great process engineering. Your process engineers need to become the key change agents in your business to deliver lean solutions. They will be the *heart* of your lean system, making parts and assemblies flow smoothly and quickly through the plant just like blood flows through our veins. Great lean process designers are able to envision and implement effective methods of lean flow, creating unique and innovative solutions that can't be bought from a catalog or off of the Internet.

Let's step back for a minute and walk through the details of why great process engineering is so important to untangling complicated production flows and, therefore, a key enabler for your lean transformation.

Your old mass production manufacturing flows were like tangled spaghetti with all of the components moving independently from machine to machine, waiting their turns in production queues. In your lean business, you are going to observe and analyze these movements to find common flows. We call these part-and-product flows *value streams*. This is exactly what the consultants will do when they walk your shop floor for the first time, but for a hefty two hundred fifty dollars an hour. Overall, you are going work to figure out ways to line up and connect the machines in these common flows so that you can link several manufacturing steps together to create fast, reliable flow. This is where the creative and innovative process engineering work starts to come in.

First you will do what can and should be accomplished through rapid change, typically through these one week kaizen events, by relocating a few existing machines to a cell to do what was once disconnected batch processing but done now in a linear fashion. For some of your value streams, your parts will flow directly through the same set of machines in an A, B, C, D fashion. These are the textbook examples where a business simply takes three or four existing machines, relocates them to a new manufacturing cell, figures out how to reduce setup times, and we are off and running with islands of lean. Great! Do these simpler lean cells and get them right. They are hugely important, but it is also rare to have all of your flows be this simple. In your business you are likely going to have only a few of these simple opportunities.

However, eventually you will also have to address your more complicated flows, those where you can't just rearrange a few machines to get the job done—flows that go A, B, $C_1/C_2/C_3$, X, D_1/D_2, and so on. What you are likely to have is a set of parts that share some common flows but then have separate flows from there, or separate spurs within common flows. The possibilities are endless; nonetheless, your business will have many complex flows in its current mass production state.

The challenge with many young lean programs is that the capability to change these more complicated processes into effective lean flows doesn't just exist somewhere "ready to serve" within the organization. Businesses start off down the kaizen path and show some early gains with their simple flows but quickly hit a wall. They find they lack the manufacturing engineering capability and resources to make machines do what they need them to do to take lean across the board. These businesses struggle in their lean journeys because they don't have the capability to create flow broadly across the shop floor, encompassing from the simplest to the most complex product lines within the business. The ability to create flow solutions becomes the early make or break for these lean journeys. The businesses get caught having some parts of the business working toward lean and some other parts where they just can't see how to get there. Lean progress often stalls at this point because the business cannot do what it needs to do with machines and processes to create flow across the board.

Successful lean businesses untangle these challenges and turn them into continuous flows through very strong process engineering capabilities. In lean businesses, manufacturing/process engineering finds ways to *cost effectively create* these lean flows through custom machines, creative setup reduction and redesign, and direct part transfers. These process engineers are the people who design specific lean solutions for your business, ones your current machines can't perform, and ones that you can't buy off the shelf. It takes a deep, creative process engineering capability to attack and eliminate big batch processes throughout most operations.

The details and makeup of this process engineering organization are covered in my Lessons 3 and 4, but for the sake of having a common terminology, think of this capability in the lean world as a *Manufacturing Systems Engineering* capability. This is a good title because it captures the essence of what these engineers do. Manufacturing Systems Engineering designs fast flow, connected machines and part transfers that work together as a system to ensure cost effective, capable, balanced lean manufacturing flows.

Creating this expert capability usually requires an upgrade or investment in additional engineering resources, an investment in creating a core competency that can attack lean process blockers and figure them out. Lean businesses need to make this investment to be able to design unique flow processes and machines as a long-term capability that sticks with the business throughout the lean transformation, enabling the business to use short cycle time, lean flows as the only way it runs the shop floor!

CREATIVE FLOW EXAMPLE

The furniture industry has so many distinct wood part flows that it is difficult to create cells with common machines directly linked together to flow parts through the factory. Traditionally, each part is treated as a unique flow with individual setups at each machine. This practice carries on in many furniture businesses still today. However, there are many opportunities where common part flows do exist and where we can design dedicated flow manufacturing cells for these through better process and machine designs.

Furniture is made up of many long, thin, solid wood parts. Walk through a furniture factory, and you will see them everywhere, thousands of, long, solid wood parts lying all over the shop floor. These are used to hold case sides and backs together, for case dividers and for table aprons and many other applications. And each of these parts is designed specifically for its own application with unique dimensions, hundreds or thousands of wood part designs that are shaped and act very similarly but differ only in variations in their dimensions.

In a mass manufacturing system, each of these parts had distinct part routings with setups at every machine for every production run even though they are processed through a set of common machines. The machines are primarily molders that make the outside shape of the parts and tenoners that cut and shape the ends of the parts. Every production run moves through the factory with the operators setting up these machines over and over again for the subtle dimensional differences in each part.

When studied, the majority of these parts fall into a few common groups:

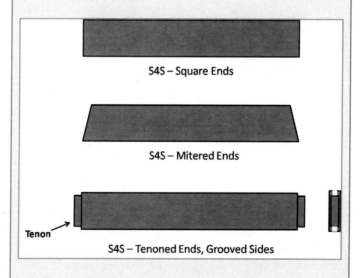

The industry uses this designation *S4S* for "surfaced four sides."

So a straightforward opportunity to lean out these part flows is to setup a cell with these common machines in a continuous-flow cell. This was called the S4S cell in one business that I worked with. The parts come into the cell from the rough mill as rough sawn wood blanks and are surfaced and shaped in this cell, therefore, the S4S, or surfaced four sides name.

In order to improve flow, this particular business decided to setup this lean processing cell. They had an extra molder that had been mothballed in one of their warehouses. The molder needed to be capable of just a few dedicated profile shapes, which could be accomplished by a finite, small number of cutter-head profiles. The molder also had a multitude of drop in heads, which could cut the grooves when required. A cutoff saw was used to cut the square and angled end cuts and to size parts to length. The company's engineers had to specially design this saw to be capable of quick, computer-controlled, automated length setups.

The old, double end tenoner that was historically used in this process took too long to setup for each of the different end cuts and part lengths, usually an hour or more. In this new cell design, the old tenoner was replaced with a new technology, single end tenoner that had 10+ change-on-the-fly heads. The change on the fly heads enabled tools for each end profile to be preloaded, and the single end tenoner could handle any part length in two passes, one cut on each end.

Also, common molder setups could be planned to run on the machines in a continuous sequence, so larger quantities of like parts could be processed without major setup changes at the molder.

The benefits of the new cell were impressive. It changed an operation that typically was scheduled for four weeks lead time in the old batch-and-queue world to one day in the lean world. The number of direct machine laborers was similar, but the material handling between machines was eliminated, saving one operator there. This savings was used to justify the new machines that the cell required. Also, this new S4S cell unburdened the old machines which could now be used to process more complex shapes. Capacity was freed up where it was needed for more complexity.

The last important benefit with this change was that the new design gave the product design engineers a standardized flow for parts that they could use going forward to standardize their product designs against. In the old world, product design engineers could make the tenons any shape and size they thought was necessary. In the new lean world, it made sense for them to stay within the bounds of the process design. This is another indirect benefit gained from this S4S cell.

What is important in this example and very different from the old batch-and-queue world is the change in process engineering demand and technique. The business needed to develop the capability to create manufacturing processes that could flow parts quickly and confidently through the plants. The furniture industry traditionally had very few true process engineers and was facing the same challenge that is the subject of this chapter, the same one that many businesses face. This business needed a stronger process engineering/manufacturing engineering capability to be able to move off square one on this S4S cell. People in the business had to design and build these machines, and they had to be cost effective. The business had tremendous lean potential but was missing a process engineering capability that was strong enough to create flow across the business.

In the case of the S4S cell, the example furniture company was able to move the ball by diverting resources to get it done. But it was clear that the business needed to develop much stronger process engineering and other supporting resources to take lean deeper into the business.

Successful lean is about getting a few critical success factors right, these enablers that you need to understand to succeed. This chapter introduced how lean is achieved through custom, unique shop floor flows across your operations, from the simplest to the most complex processes. Lean businesses need to develop a passion for great process design. This stronger process engineering capability has been one of the most important lean success factors in my experience. Businesses that cannot develop flow out of their mass production chaos will not become lean, period. And the way we actually do this is through this model of lean Manufacturing Systems Engineering; a strong, creative process engineering capability. This is a non-negotiable enabler of great lean manufacturing.

So now that we know that continuous flow production across the shop floor is the only way we can create a lean business, our next step becomes how do we actually create the organization that does the work? I cover the details of this Manufacturing Systems Engineering model in the next two lessons.

LESSON 3:

LEAN "MANUFACTURING SYSTEMS ENGINEERING" CREATES FLOW

A strong manufacturing engineering capability creates cost-effective lean processes that flow products quickly and reliably through the plant.

I once supported a group of individuals that was working to drive a corporate wide lean initiative. This particular business had a long history of batch-and-queue manufacturing and huge levels of work in process and final goods inventory. The business had tremendous potential to lean itself out and was ripe with areas where it could start to put machines together to create flow, but it also had a critical blocker—it didn't have a strong process engineering capability. The business had a traditional machine shop with fine capabilities, and it had a handful of corporate and divisional process engineers, but the depth and agility of those were not strong enough to drive lean successfully across the business. The machine shop was unionized and working through traditional paperwork processes with work orders and structured lead times, and the process-engineering group was too small and well trained in the ways of batch processing; buy a machine from a catalog, put it on the shop floor, get it working, and move on to the next project.

In lean businesses, process engineers work hand in hand with the product engineers and operations to develop *every* manufacturing process into a lean flow. Lean is commitment to never putting another batch process on the shop floor. This business's weak process engineering was keeping it from achieving lean success. The business needed to develop a much stronger process engineering capability with responsive resources and the creative capability to design lean flows. It needed to change its process engineering into a high performance team, focused and capable of only creating lean process going forward!

In the end, the business didn't have the leadership vision or belief that stronger process engineering was one of its keys to succeeding with lean. The leadership couldn't see the benefits of changing its resource allocations to create this potentially profitable new capability. Over time the corporate lean team worked to build a small skunk-works organization that could do some lean machine design activities, but they were relegated to build this capability one brick at a time, under the radar screen. What was accomplished was meaningful, just too slow and small compared to the opportunities. This lack of a strong manufacturing engineering capability was the governor on the business's lean engine. It was keeping the business from making a broad conversion to lean manufacturing.

As discussed in the previous chapter, every successful lean business is *created* through the capability to design unique, custom, continuous shop floor flows. Lean manufacturing is fundamentally about creating faster and more efficient production flow by linking machines together in continuous flow lines where parts move quickly from machine to machine with short setup times and little inventory. In order to actually do this, we need to retool our traditional batch-and-queue process engineering approach to create what successful lean businesses have called a Manufacturing Systems Engineering approach. This Manufacturing Systems Engineering organization consists of process engineers and other related resources that do the work of designing and implementing lean manufacturing systems. (I will use the terms "manufacturing engineering" and "process engineering" interchangeably throughout the book, as many businesses do.)

This lesson covers the details of what makes manufacturing engineering so different in a lean business such that it requires building a much stronger capability than what was required in mass production. If we can understand what these resources actually do and how they do it differently in lean manufacturing, then we can start to get our heads around how drastically changed process engineering is in a lean business. We can then use that knowledge to develop a *model* of how to organize and create this capability in the next chapter.

So what does Manufacturing Systems Engineering actually do that is radically different in effective lean businesses? Remember again the key enablers of flow from our lean principles:

> ## LEAN PRINCIPLES
> ### Lean How-Tos:
>
> - We identify common part and assembly flows, or "value streams."
>
> - Within these value streams, we design long machine setups out; we design new machines/processes with little or no setup time so we can manufacture small batches cost effectively.
>
> - We link these machines together in new continuous flow value streams that replace our old disconnected value streams with cost-effective, small-batch, continuous flow production.
>
> - We move parts immediately from machine to machine, with little or no work-in-process inventory so that products move rapidly through manufacturing, with short processing lead times.
>
> - Theses short lead times enable us to manufacture parts and assemblies to actual demand or closer to actual demand, not long range production forecasts.

This is exactly the work that Manufacturing Systems Engineering does to create a lean business. Complex manufacturing businesses need creative, disciplined solutions to create smooth, continuous flow out of historically chaotic part flows. This is one of the key capabilities that you will need to develop to enable lean success across the entire business. One of the greatest shortcomings, if not *the* greatest shortcoming in weak lean programs is the inability to create these custom shop floor flows. Restating: making lean work in your business is really going to come down to how well you understand and *do* the bullets above. Lean businesses develop a capable Manufacturing Systems Engineering organization to do exactly this work!

Following this concept further, next are what I believe are the four fundamentally different challenges that you Manufacturing Systems Engineering (MSE) organization will face in a lean environment:

1. Manufacturing Systems Engineering works to eliminate setups and link machines to create flow everywhere on the shop floor.

2. Manufacturing Systems Engineering works hand in hand with the product teams to develop lean solutions for every product flow, all existing products, all new products, and also for productivity projects.

3. Manufacturing Systems Engineering designs and builds custom machines to achieve lean flow.

4. Manufacturing Systems Engineering must be empowered and agile to support rapid, continuous improvement.

Let's explain each one of these items in some detail.

Manufacturing Systems Engineering Works To Eliminate Setups and Link Machines To Create Flow Everywhere on the Shop Floor.

In mass production, manufacturing engineers learned to work in the following way. They searched for capable machines with high throughput, reliable quality, and relatively short setup times. These short setup times were desirable but still not deemed critical in mass production. Mass production engineers were smart enough to understand that less setup time was valuable, yet the engineers were permitted to buy and install machines that had some level of setup time because batch-and-queue mass production was built around those.

And as was state of the art in that system, the machines went on the shop floor with groups of like machines. They were hooked up to the necessary electrical and plumbing systems, qualified, and then released to manufacturing. And, yes, there was a high degree of innovation in that system to create better ways to process parts and assemblies, just not focused on the key variables that a true lean system requires.

That, in short, was the role of a traditional process engineer in, let's say, the 1970s.

So what is different today in lean Manufacturing Systems Engineering? First of all, in lean manufacturing, wherever possible, you are going to take out setups completely or make them very, very fast. Setups mean you have to wait between processes. Wait time means you can't move parts continuously through, machine by machine, and flow parts quickly through the shop floor. Long setups mean we have to run large batches of parts through each setup to amortize the cost of the setup. Put a handful of these long setup machines in a manufacturing flow, and we go right back to big batches. It's the only way to process cost effectively with long, expensive setups. A small number of long setup machines anywhere in the direct production flow directly results in big batches, which in turn results in longer lead times, and, in turn again, we've just killed any chance for direct, lean flow.

Think about this for a moment. As a simple example, if we have even a very short setup time, say a five-minute setup, with a fifteen-second-per-part run time, which is typical, what size(s) batch will we end up wanting to run to be cost effective? We can't run one part at a time because that would kill the business with all the cost of the setup going into just one part. And nothing would get through the factory because all we would be doing is setting up the machines all the time. If we run twenty parts through the machine, we would have five minutes of run time and five minutes of setup. That's not very cost effective either but plausible. We are normally going to want to run at least one hundred to two hundred of these parts per setup to be cost effective and have high throughput.

Bingo. Right there we've created the need for batch manufacturing! That's how it works with setups and fundamentally why mass production ended up as batch-and-queue manufacturing. Allow what seemed like short setups in a mass production environment at even a handful of machines in your factory, 5 minutes in my example, and you are quickly going to be relegated to batch and queue manufacturing across the plant floor. Once any one machine in a given flow needs to be a batch processor, all the other machines in that same flow are almost always forced to process that same batch too. We can't practically unbundle and re-bundle large batches as they move from long setup machines to short setup machines; once a big batch, always a big batch on the shop floor. Simply stated, medium and long setups will not allow us to achieve lean manufacturing!

However, and as a point of understanding and for the sake of completeness, there are viable exceptions to this "no long setup" rule in lean manufacturing systems. On standalone machines with excess capacity and little use, we can sometimes tolerate relatively longer setups. Since the machine is sitting idle most of the time, and it's not linked to the next process, we can theoretically take care of setups while the machine is not being used. Another example of where setups can be tolerated is to coordinate setups across a family of machines to tolerate longer setups somewhat cost effectively. But in nearly all cases setups are the enemy of lean and continuous flow, so we work to take them out of the process wherever possible. I hope you get the picture; setups result in batches moving through the factory, even with reasonably short setup times. Long setups, because of their resulting large batches and inventory, are the direct enemy of lean!

So at the core, this lean manufacturing change is really a mind shift in the way we approach process design. To be truly successful with lean, your manufacturing engineering must design processes that flow linearly, quickly and directly from machine to machine. In lean we group these machines together in cells so we can process multiple steps of the manufacturing process in a continuous flow. To do this you must design processes with little or no setup time and part transfers that are smooth and continuous. Therefore, you are going to have to take setups out of nearly all of your machines to make parts and assemblies flow cost effectively and benefit from really short lead times.

This is where the concept of *one-piece flow* became an important directional goal in lean. If we can cost effectively design a group of flow machines that can move mixed parts through one at a time, we have designed the perfect production system! With zero setup time across the board, in theory we can cost effectively build one of anything because we've designed out the setup costs. One-piece flow is a valuable goal: it enables us to build exactly what we need only when we need it!

Manufacturing Systems Engineering Works Hand in Hand with the Product Teams to Develop Lean Solutions for Every Product Flow, All Existing Products, All New Products, and also for Productivity Projects.

In lean businesses, process engineers work tirelessly with the product teams to design flow processes, and they do this universally. As time goes on, every process on your manufacturing shop floor will become part of a lean flow. Eventually everything you do newly on the shop floor is also designed to be lean. Your process engineers are going to work with your product engineers to design out all of the glitches that cause batch-and-queue production. They are also going to eliminate long setup machines because these are batch machines. You simply are no longer going to allow long setup times or long cycle times in your business. Eliminating those is the way only you can achieve lean across your operations!

Your process engineering organization will work together with the product teams to produce lean solutions for every major manufacturing activity in the business. Initially they will use it to clean up the existing shop floor and convert the business to lean. Process engineers will also be involved in your productivity products to come up with more cost-effective product and process designs. When you replace outdated equipment and install new capacity, it will need to be designed to be lean by your manufacturing engineering group. And to keep the business lean forever, every new product will be designed with lean process flows.

Many of these improvements will be developed through the various types of one-week kaizen continuous improvement events. The kaizen process is universally used in lean a way to conceptualize incremental process improvements across business activities and rapidly put the improvements in place. For example, *Shop Floor Kaizen Breakthrough* (SKBs) events are used to develop radical productivity improvements for existing manufacturing processes and to implement them within these one-week kaizen event. The process pushes the organization to make improvements rapidly, no waiting around for eight to twelve weeks to rearrange and modify a set of machines.

Your process engineers will be involved in all of the kaizen events that affect shop floor flow, and your lean Manufacturing Systems Engineering capability is going to be the heart of these process improvement activities, pumping blood through the business in the form of continuous product flows. This is the on-call, hard work that it takes to create and support a lean manufacturing enterprise. No rest for the weary process engineers!

Manufacturing Systems Engineering Designs and Builds Custom Machines To Achieve Lean Flow.

Your lean process engineers have to work relentlessly to take setup and long processing times out to create continuous-flow processes. The core task of your lean approach is to identify common flows and then design cells that put those machines together, taking setup out of each process in the value stream so parts and products can flow step by step in short cycle times. These solutions are almost always unique to your business. Your culture is unique, your products are unique, your plants are unique, and your lean-flow solutions are going to be unique too!

In the lean world, we need process engineers that can use their brains to identify and eliminate process flow blockers. Businesses find out early in their lean journey that they can't buy these processes out of a catalog. These engineers will need to design process solutions that are unique and custom to your product lines.

Lean flow cannot be created by buying a machine from vendor, finding a place on the shop floor for it, hooking up the power, compressed air and dust collection, and turning it on. This is what the lean consultants like to call "catalogue engineering" and that was good enough for mass production. In the lean world, you are going to have to throw out catalog engineering and become very good at designing your own manufacturing machines and processes. Custom machines and modified machines are the building blocks that will enable you to create lean, continuous-flow processes for your business. This is the only way to create truly sustainable lean.

Manufacturing Systems Engineering Must Be Empowered and Agile To Support Rapid, Continuous Improvement.

Following is the schedule for a typical five-day shop floor kaizen breakthrough event:

Day 1: Train

Day 2–3: Analyze, Design, Decide

Day 3–4: Rearrange, Build, Rebuild

Day 4–5: Practice, Run, Standardize, and Celebrate

Day 6+: Follow up

Is your batch-and-queue manufacturing engineering capable of doing work this way, designing new processes, building them, and then actually installing them for use in production in less within one week? No way! Those folks are used to being in a nice, cozy office most of the time where they drink their warm coffee all day and don't get their hands dirty. Further, in your current organization, there's nobody behind them to do the physical work of rearranging and modifying machines in this short time frame. Try to get a unionized, work-order based skilled trades organization to rearrange, build, repair, and setup a new assembly line in two or three days. Good luck with that in your old mass production structure!

Again, you are going to need to retool your manufacturing engineering capability to become lean. Your lean process engineers will need quick decision making authority and a budget to get work done rapidly. They are going to need space and rapid prototype tools to mock up their ideas and try them out. They are going to need management available and attentive to support quick capital investment decisions to choose production directions and run with them. Bureaucracy and stodgy structure get in the way of making the decisions that drive lean forward.

So those are four fundamental changes that your process engineering organization will face in a lean environment. Think of these four differences in a lean environment in the following simple way: *your manufacturing engineering organization is going to have a heavy workload; they are going to need to be capable of designing unique, creative machine solutions to enable flow; and the solutions will need to be done efficiently and quickly to support continuous improvement*

So what is the "so what" with this change in process engineering? The difference in lean is that your manufacturing engineering is going to have to be more involved, more capable, and more innovative to create flow across all of your manufacturing operations. As we learned in Lesson 2 and 3, this isn't going to be as easy as just putting existing machines together and magically creating instant flow. Great process engineering is going to be necessary to make lean flows possible where it wasn't before through the creativity and invention of new machinery and processes. It involves creating flow as a system, with synchronized machines, short setups, and seamless part transfers. This is why lean experts added the term "systems" to *Manufacturing Systems Engineering*. Also, your process engineers are going to have to have a system of resources alongside them to take these ideas and turn them into real machines and flow lines, people who fabricate these designs into actual machines. Then we need people to install the machines, get them running, and maintain them.

Manufacturing Systems Engineering designs connected machines and part transfers that work together as a *system* to ensure cost-effective, capable, high-quality part flows. And, building this comprehensive manufacturing engineering capability is the number one fundamental challenge to creating successful lean businesses. Businesses that don't invest in and cultivate these specific process engineering capabilities are not strong lean manufacturers. The bottom line is you will need to dedicate substantial, agile, creative resources to design, build, and install your own custom machines, jigs, fixtures and part movements to create real lean flow. Period!

So now that we know that manufacturing engineering is so different in the lean world compared to the mass production world, our next challenge becomes how do we structure and staff this team? Let's cover the details of this Manufacturing Systems Engineering operating model in my next lesson.

LESSON 4:

THE LEAN MANUFACTURING SYSTEMS ENGINEERING MODEL

Successful lean businesses create their own Manufacturing Systems Engineering expertise, which has separate and defined capabilities to manage, design, build, install, and maintain flow manufacturing processes.

You probably have an organization in place today that wears the manufacturing engineering hat; it is usually called manufacturing engineering or process engineering or something similar to that. Your lean approach is going to bear similarities to that existing capability but will be stronger and structured differently to be capable of creating your lean-flow processes. No more catalog engineering, no more big batch machines, no more disconnected machining centers, and no more slow responses to today's manufacturing needs!

As was introduced in Lessons 2 and 3, I want you to think of this capability in the lean world as a *Manufacturing Systems Engineering* (MSE) capability. Again, this is a good title because it is precisely what this team of people does. Manufacturing Systems Engineering designs connected machines and part transfers that work together as a system to ensure cost-effective, capable, high-quality, part flow.

So let's puts some depth into how this new type of process engineering capability looks and behaves. You can follow the model designed here to develop your own specific organizational structure

There are five key distinct capabilities that make up a successful Manufacturing Systems Engineering capability in a lean environment:

- Process Engineering and Project Management
- Machine Design
- Machine Build
- Machine Installation
- Preventative Maintenance

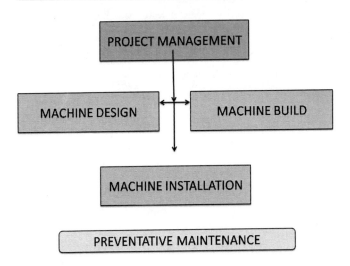

Following is a short description of how each of these functions contributes to a complete and capable lean Manufacturing Systems Engineering core competency.

Process Engineering and Project Management: Within MSE a group of manufacturing engineers develop the processes and high-level system designs for all of your lean process flows. Importantly, they also do the overall project management coordination to deliver successful lean projects. Think of this group of people as the overall systems level designers and project managers for all manufacturing that takes place on the shop floor. This group is made up of primarily mechanical and electrical engineers with supporting technicians for detailed design activities and shop floor work. These individuals are the project coordinators; they work directly with internal customers, usually the divisional product engineers, to translate product designs into process requirements. They attend the divisional product planning activities and the kaizen events to understand the new product and process needs and then develop these concepts into lean process flows. They evaluate options and make decisions on the final designs for the overall production system. They work with the machine designers (below) to translate process needs into actual machines.

Machine Design: These are your machine design experts. They take the concepts from the manufacturing engineers and turn them into actual machine designs. Most of the true creativity in lean process design comes from this specific group of people. The group consists of your most creative design engineers, the engineers who like to tear down and fix old farm tractors on the weekends. The Machine Design group also contains CAD designers who do the detailed machine designs. This team is where the rubber meets the road, where good product ideas are turned into can dos or can't dos.

This machine design capability is one that you can build internally, or one you can outsource. In the early days of your lean conversion, it may make sense to have a local engineering design shop to help you get started. Eventually, however, and in my experience, you will benefit by building a machine design and build capability internally as is discussed in detail below.

Machine Build: Machine build consists of two separate functions. The first group is your skilled tradesmen. They are machinists, welders, painters, carpenters, assemblers and the like who make the parts that make up a machine. You probably have a skilled-trades area today in your business. You can use these same resources to build your lean manufacturing machines as long as you build a culture of responsiveness so they can work in short time frames to support your lean process designs and continuous improvement activities.

The second group within the machine build area is your machine assembly group. These are the people who assemble machines and get them working, qualified, and ready for production. They work with the machine designers to get everything working just right.

You likely did not have many people dedicated to the art of building machines in the past. This capability is one you probably need to create newly to enable your lean business. Again, you can outsource this capability, and, for that matter, you can outsource the entire activity of machine design and build. However, this is an important decision point, as the machine design and build requirement can be demanding in a lean environment. There is continuous and intense interaction across the project management, machine design, and machine build teams. As this activity grows into a core part of your lean efforts, you are likely going to need to have it close at hand to run the lean playbook on an ongoing basis. I cover these pros and cons of out sourcing in deeper detail below.

Machine Installation: You are going to need to have a group of hard-working, flexible people who can install and commission your lean processes. The machine installation crew is a focused team that moves, rearranges, and installs new manufacturing lines and machines wherever and whenever they are needed in the business. This is a multi-disciplined team with all of the skills needed to install machines in your business; electricians, pipe fitters, welders, and assemblers to name a few.

This team installs the new machining cells and processing lines that come out of your process design and machine design/machine build activities and get them up and running. When you get good at lean, you will have a couple of lean process improvement (kaizen) events going on in the business every week. The machine installation crew will also support these continuous improvement (CI) activities so the process improvements can be implemented in rapid fashion. It is not unusual to completely change the layout of an assembly cell in these one-week kaizen events. The machine installation team is *the* group of individuals that do the heavy lifting to make these changes, often working around the clock to get things ready for production the next day. We can't be without our production equipment for more than a few hours. In lean, our lead times are so short, and the rubber band is stretched so tight, that losing any production time causes an immediate disruption in order fulfillment. This machine installation team is critical to having an effective, ongoing, continuous improvement capability.

The machine installation team spends their time where the CI activities and new process installations need them. Your lean program will put high demands on this group. In fact, there will often be more demand than resources here, as each of the plants and divisions works to deliver their annual productivity targets, product improvements and new product launches. In multi-site businesses, this team normally spends 50 percent or more of their time traveling to support every operation. This is an area that requires some creative management and labor agreements to handle all of the travel, late hours, agility needed, and just plain hard work. Because of this intense work demand, creating this capability will take some focused work within a union shop to craft satisfactory labor agreements that can match a lean business's needs.

Preventative Maintenance: The last key process capability that you will need within your Manufacturing Systems Engineering organization is an ongoing preventative maintenance function. Lean businesses must keep their processes running every minute of every day. This was obviously also important in mass production systems, so you might be wondering why the difference in lean? In lean, you are going to have machines connected together to create continuous flow throughout the factory. With the short lead times we drive for and create in lean, and with little inventory in the system, when a machine goes down anywhere in the factory, output crashes very quickly. We have processes and people sitting idle very quickly when we have an abnormality in lean's continuous manufacturing cells. And we very quickly, possibly even instantaneously, start to affect output from a lean factory when process breakdowns occur. In mass production we had heaps of inventories to help cover up these machine breakdowns and keep some processes running while others were idled by machine problems. So mass production could keep producing in some areas for some time even when individual processes went down. The point is that preventative maintenance is critical in both lean and mass production however the sense of urgency is increased significantly in a lean manufacturing system. In lean we take all of the work in process inventory out of the system and therefore break downs impact throughput nearly immediately.

You will need to have both preventative and reactive maintenance processes and maintenance people in place to keep your lean processes running well, nearly every minute of every day. This is the only way to consistently deliver lean's continuous flow, short production, lead times. As such, preventative maintenance becomes a critical part of successful lean manufacturing systems. The term "total productive maintenance" (TPM) has come to be associated with this lean requirement since maintenance in a lean environment is focused on continual, ongoing maintenance to keep throughput and productivity high.

What is the scale and scope of a lean preventative maintenance organization? It's difficult to be prescriptive here since every business figures this out a little differently, but there are usually three basic parts to a solid preventative maintenance strategy:

1. Daily preventative maintenance
2. Periodic preventative maintenance performed by maintenance experts
3. Reactionary machine support to fix breakdowns or other abnormalities

The first of these requirements is that you assign ongoing preventative maintenance to your direct labor as part of their standard operating procedures (SOPs). These maintenance activities include the more simple measures that need to be done on a daily basis; keeping the area organized and clean (5S), lubricating and oiling machines, checking gauges and poke yokes for accurate operations, and other maintenance activities that keep machines ready on a shift-by-shift basis.

The second and third of these activities are usually performed together by one central maintenance group. This team is staffed to take on these larger regular maintenance activities and is also available to fix machines when they break.

Periodic preventative maintenance requirements are determined through the machine manufacturer's recommendations and through experience running them. These regular periodic maintenance activities include major machine lubrications, replacement of wear parts like bearings, welding tips, weld plates, wear guides, motor replacements and a host of others. Preventative maintenance often also includes the use of periodic infrared scans to identify temperature buildups in motors, bearings, transmissions, and other moving parts, to find parts which are wearing out and need to be replaced. Vibration detection can also be used to find moving parts that are running erratically and may also need to be replaced.

These same central maintenance people are also on call to react immediately to fix machine breakdowns when they occur. As a part of your preventative maintenance strategy, you will need to stock an inventory of replacement parts to keep your machines running when these breakdowns occur.

You probably already have a group of central maintenance people. The lean journey becomes an effort of evolving them to have the right sense of urgency and processes to keep your lean processes running all of the time.

These are the five key functions that in lean make up a successful Manufacturing Systems Engineering core competency. Again, the most successful lean businesses grow this Manufacturing Systems Engineering expertise internally so that they are able to create, manage, and evolve their own continuous flow solutions. As developed in this and the last two chapters, creating this capability is one of *the* most important organizational changes that will enable you to develop lean manufacturing solutions for your business. It takes a deliberate management decision and a change in course to realign manufacturing engineering resources so that they are dedicated to creating flow—resources that normally do not exist in a mass production organization in the right structure, scale, and with these new lean intentions. However, when done well and according to my model, this team will be *the* key enabler in your lean business, capable of developing and implementing lean processes across your operations.

There are a few other important supporting aspects of growing and managing this new lean Manufacturing Systems Engineering capability that we need to cover, as follows. First, and as a word of caution, many businesses get derailed on their lean journeys because they don't grasp the need to commit to building this Manufacturing Systems Engineering core competency *early* on in the lean journey. Great process engineering becomes a big resource commitment and demand right out of the blocks in the lean transformation process. Your CI efforts are going to uncover needs for new, custom machine and flow capabilities very early on. The capability to solve these did not exist strongly in your business before lean, and the decision to make significant, early organizational change can challenging for businesses that haven't done it the lean way before. The process requires management direction and support for Manufacturing Systems Engineering very early in the lean journey.

Further, you are going to find that it is one thing to envision lean flow on the shop flop and another to be able to actually make it work. The challenge is to figure out who is going to do all of this new work and then empower them to do it. Therefore, the organization that designs your manufacturing systems is going to need be empowered with agile decision-making capabilities and access to funds to be adept at creating custom process designs for your business. This may also require a change in the culture of how you go about decision making for process design and execution. We have to drive decision making to the lowest possible level in this respect and empower these resources.

Manufacturing Systems Engineering is a leading capability that enables your lean initiatives to get out of the blocks successfully, one that has to have strategic leadership support early on. In other words, creating this Manufacturing Systems Engineering capability becomes and early inflexion point to the success of your lean changeover. If you can't swallow this change and make a substantial commitment to it, your lean implementation will likely not be successful very deep into your business. These early needs will get in the way of lean flow if you don't have a way to solve them with great process engineering. However, when you do embrace this change, the Manufacturing Systems Engineering model becomes one of the keys to developing lean into a competitive weapon; enabling your business to create long-term, sustainable lean processes across the shop floor.

The second important supporting aspect is about pace. Specifically, how fast do you go with building this lean manufacturing engineering organization? You are obviously not going to develop this competency overnight like turning on a light switch. You are also not going to kick this off as your first initiative to drive lean and throw dozens of resources into it, hoping that they get it right. So, you need to build this competency alongside the needs and pace of your lean initiatives on the shop floor. Start small with a dedicated team that you carve out of your existing resources and then build it as you monitor the progress of your lean changeover. You will hit many speed bumps in the early part of your journey, but you must continue to massage these organizational capabilities to drive forward. You need an overall sense as to where you are headed and commit to it and then build the capabilities over time. It will take years of designing your own processes to get really good at it, for your process engineers to understand how to effectively use this new enabling capability to design world-class manufacturing lines. You are going to build this capability alongside the pace and scale of your lean progress.

My third key supporting aspect of successful Manufacturing Systems Engineering is this decision of outsourcing versus in-sourcing the machine design and build capabilities. Most mass producers are used to buying machines off the shelf. You are in the business of making products not machines. Machines you've needed in the past have been readily available from experts that do machines as their business, so why would you want to take on designing and building your own machines? And your machine demand has not been constant enough to build this expertise in house and keep it busy. All this makes rationale sense, why would you want to become a machine expert as a part of the lean transformation?

Let's peel this onion back a little more. You have an important, strategic decision to make in lean regarding how much of this machine design and build work you do internally and how much you farm out to outside shops. But first let's make sure we are on the same page about what kinds of machines this group will design and build. You are not going to start designing your own two-hundred-ton presses or your own robot arms internally through this machine design and build capability. Those highly complex machines will still need to be purchased from machine specialty companies even in the lean world. Still, your flow lines are going to require many specialized work stations, punches, jigs, nests, transfer mechanisms, and poke-yokes that will be unique to your business. Lesson 5 covers several of the tools manufacturing engineering can use to attack process problems and create flow through these types of machines. You will see in those tools an endless opportunity for simple, realizable custom machine and processing aids that can help lean factories flow. These are the machines where you will start to build your internal machine design/build capability.

So the question is do we take on machine design/machine build in-house or source it on the outside? Having expert engineers in-house who understand what machines can and cannot effectively do is a huge enabler for your continuous improvement efforts. Very few of your lean processing challenges will be able to be solved with catalog machines. They are going to require unique, creative, cost-effective solutions. Further, you will uncover solutions well into your lean journey that you could not see in the early days. You are going to have many "aha" moments as you implement lean— "If I could just do this or that with my machines, I could take time out, or labor out, or improve quality or all of the above." These ideas are the heart of continuous improvement, and their strong potential benefits will go untapped without the capability to develop custom manufacturing solutions. However, when you do commit to an internal machine design and build capability, your capabilities will grow along with your lean efforts, generating effective, proprietary solutions for the business.

Further, you are rarely going to get custom machine designs perfect the first time. It is the nature of creating and inventing new ways to get things done. You are going to see things in your machines that you want to change to make them more robust after using them for some time. Once your continuous-improvement minded people use the machines, they will bring forward ideas to make your machines even better. When you create this machine design and build capability internally, it is there to be responsive when you want to improve your machine designs and to fix things that go wrong with your machines—it is more agile and capable of making these continuous improvements.

On the contrary, when you create this machine design/build capability on the outside, your team does not develop the culture, the capability, and the collaborative spirit to continuously do things better. When something needs to be changed or improved on your machines, you are tied to the costs and responsiveness of outside experts instead of an in-house staff. These custom solutions are slow and expensive when outside suppliers do most or all of the work. When machine design and build is done on the outside, it simply becomes a blocker to continuous improvement.

The most successful lean companies create expert machine design and machine build capabilities internally. An internal machine design and machine build capability helps to create an entrepreneurial can-do spirit that fuels the continuous improvement efforts of the business. This is a strategic direction you need to think through for your business; in businesses that decide to fully commit to lean, creating this capability internally is usually the better decision.

In summary, your business was built around innovative and unique products, and that is what makes them desirable to your customers. Your processes have also always been unique not only because your products require it but also because you've had innovative and proprietary developments on the shop floor throughout your company's history. Lean is going to push this need for unique processes even further. Great process capability, a Manufacturing Systems Engineering type of capability, is the proven way to create custom, continuous flow processes that make manufacturing operations truly lean. This is where the true core of the lean change occurs!

LESSON 5:

TEN LEAN FLOW TOOLS

> Use proven manufacturing engineering approaches to create smooth, linked lean flow.

The purpose of your Manufacturing Systems Engineering organization is to eliminate all process-flow blocks to creating lean flows. Their purpose is to design all processes on the shop floor so that they flow smoothly and directly without any stoppages for long setups, inventory queues, or other processing delays. However, there is no formula for this. You cannot buy this capability; you simply must grow it. It comes down to talented resources, an open environment for creativity, brainstorming and planning your projects with an uncompromising expectation for lean solutions. Fortunately, there are several tools and techniques that manufacturing engineers can learn to use to attack process blocks and create flow. Following are ten effective tools that process engineers use to create successful lean flows:

1. Dedicated, single-purpose machines
2. Right-sized machines
3. Flexible machines
4. Data-driven processes
5. On-the-fly setups
6. Poke-yokes
7. Chaku-Chaku cells
8. Paced production flow
9. Automated part transfers
10. Combining machines

1. *Dedicated, single-purpose machines:* Isolate a repetitive group of processes in your factory and reconfigure these into a dedicated cell with the machines placed in a focused manufacturing flow cell. Dedicate each machine in the cell to a singular activity, drilling and tapping one hole, for example. With dedicated, single-purpose machines, you never have to stop the machines to do part setups; you only need to maintain them and keep them running well. Parts will fly through dedicated machines with very short, consistent lead times.

Example: A kitchen cabinet factory may have as many as ten or fifteen edge profiles for doors/drawer fronts. Instead of molding these profiles on a double-edge tenoner, a huge machine with long setup times, dedicate shapers to make each profile. Shapers are inexpensive, roughly one-twentieth the cost of a tenoner. For your low-volume profiles, you can use one dedicated shaper that a worker manually feeds each of the four sides of a door through. For high-volume profiles, you could place four shapers in a cell and outfit them with load/unload feeders so that the operator moves through the cell loading and unloading a door into each shaper, turning them ninety degrees with each movement. That way each shaper is dedicated to molding one edge of the door. This four-shaper cell will generate significantly more output per time compared to the single-shaper cell.

2. *Right-sized machines:* Machines should be about the size of the part they are machining. This is a good rule of thumb for cost-effective machines and for floor space utilization. Remember the old TV cartoon with a giant machine whittling down a huge tree trunk to make one toothpick out of it? That visual captures the essence of why we right size machines.

    Example 1: The dedicated shapers in the previous example are a good example of right-sized machines. The shaper is about the size of a cabinet door. A large multipurpose tenoner is too big, several times larger than the part it is processing. A much stronger cost/benefit can be achieved by these right-size/right-cost, dedicated machines.

    Example 2: Small presses dedicated to a part processing line to make holes in plastic or metal parts instead of giant, multipurpose centralized presses is another good example.

3. *Flexible machines:* Use machines to do multiple operations through quick/automatic changeover tooling.

   Example: Use a single edge tenoner with automated tool changes to mold all of the different edge profiles on your kitchen cabinet doors. These machines do the profile tool changeovers automatically in seconds, creating the flexibility to machine ten or more profiles on a single machine.

4. *Data-driven processes:* A really great way to enable your machines and labor to be more cost effective is to use you order entry data to perform machine setups automatically. This is one of the best tools for businesses with configurable product sizes like building products—kitchen cabinets, windows, doors, and the like. These building products come in so many different sizes that parts and final assemblies are usually built to order. So when the orders come in, use the order specifications to talk to your design data to determine the exact part sizes you need to make at your cut down machines. Then electronically transfer that data to your machines to tell them exactly what to make in sequence. This can be a huge improvement over humans trying to enter all of this data directly into each machine.

   Example: You decide to make kitchen cabinet doors with mitered stiles and rails. You stock these profiles in ten-foot lineals that you cut down to size and assemble to order. Use incoming order data to tell the cut-down machines what sizes they need to cut and have moveable stops that automatically setup to cut these sizes. You can even use the data to "look ahead" and run some math on the incoming orders requirements to optimize cut sequences and to minimize the scrap from your ten foot lineals. This is a powerful lean design tool with almost zero setup time, lower scrap, and it can help to move in the direction of a build-to-order shop floor since parts can processed automatically from orders.

5. *On-the-fly setups:* Have your machines adjust for different part length dimensions automatically. This is a common solution for cutting down extruded vinyl and aluminum lineals into multiple lengths. It can also be used for woodworking industries that need to cut parts down into multiple lengths. You will achieve even greater benefits if you can set the lengths automatically on the machine by using the design data in your computer system.

Example 1: Cutting down long parts like extruded parts or wood components in the furniture industry. Set up a saw with a moving stop to automatically setup for length. There are commercially available versions of these

Example 2: All wood factories generate a ton of leftover pieces that are too short to use in their aesthetic components. You can finger joint this scrap to make parts for your non-cosmetic, structural stringers that are used inside case goods. This cell has three parts. The first part is a set of machines that can finger join these pieces back into long lengths. This is a common process that you can buy on the shelf. The second part is to mold this reclaimed finger jointed wood into a standard width and height to be used across your product lines, say 3/4-by 2 1/2 inches. The third is to design a double-end tenoner with a standard profile for the end cuts, but one that can change over on the fly to different length dimensions, like a tenoner with one stationary head and one moving head. You can put these three machines in one cell and create all of your interior parts as you need them. The cell can use scrap to make finger jointed parts and cut them to length with no setups. It will save big money by using scrap for structural components and will also give you a nice "green" marketing claim!

6. *Poke-yokes:* Poke-yoke means "mistake proofing." Poke-yokes are error proofing devices that prevent a defect from going to the next step in a process. They are not really machines in the same sense of these other lean tools, but they are an important component of your machine design activity to ensure quality. Poke-yokes can take many forms, such as go/no-go gauges, indicator lights, movement detectors, and so on. A light beam might be used to make sure an operator picks the right part in an assembly sequence.

There are a multitude of creative ways to create poke-yokes; the list is nearly endless. Many of the solutions come from quality improvement projects that seek ways to mistake proof repetitive manufacturing defects.

Example 1: Parts are shipped in a small bag for the customer to use during assembly. Use weigh counting via an accurate scale as the parts are placed in the bag to ensure everything needed is present.

Example 2: Cabinet hardware choices allow for several grades of hinges. Use pick-lights to indicate which hinge(s) to pick for a particular cabinet based off of a bar code scan of that particular cabinet order tracking information.

7. *Chaku-chaku cells: Chaku-chaku* literally means "load-load" in Japanese. This is a powerful method of processing parts through several machines using a single operator. The idea is to setup several machines in a counterclockwise U-shaped cell and then to have an operator move the part from machine to machine, loading the parts into each machine where a portion of the manufacturing process is performed at each individual machine. Counter-clockwise movement has been shown to be a more efficient parts movement method in Industrial Engineering studies. Chaku-chaku is a good method to break up machining into several smaller discrete steps so that we can have zero setups, create one-piece flow, and use dedicated, lower cost machines for each step.

The machines used in a chaku-chaku cell are usually simple machines to keep the capital investment for these dedicated cells low. The machines are also often dedicated to one product flow because the machines are not designed to handle multiple shapes and part changeovers. Chaku-chaku is a way to use low-cost, simple single purpose machines, linked together in a step-by-step fashion to create flow for low to medium volume parts. Since the machines are single purpose, you can adapt low-cost, on-the-shelf machines like mills, lathes, and drill presses to do the work.

Example: Machining a brass cast body for a faucet. You might have one machine bore the waterway, one tap the threads, another grind the mounting surface, and a couple more to rough machine the faucet body to get it ready for final hand polishing. In traditional functional manufacturing that would take at least four or five setups across a similar number of machines. The costs of these setups and all of the non-value added material handling could be eliminated in a chaku-chaku cell, with one machine dedicated to each machining process.

8. *Paced production flow: Takt time* is defined as the work time per part at each workstation in a cell. The way we determine takt time is by taking the daily work time available and dividing it by the demand rate over that same time period. So if we have seven hours (four hundred twenty minutes) of work time available in a day and the demand rate over a day is two hundred ten parts, we have a takt time of two minutes per part (four hundred twenty minutes/two hundred ten parts).

Many well-designed lean cells use manual transfers to move parts from one workstation to the next. Automated part transfers via conveyers or other means can be very expensive and require high volumes to justify. The challenge with manual part transfers becomes keeping each station performing consistently at takt time without having an external driver like an automated conveyor or assembly line. For example when one operator drops a needed screw and misses his or her takt time, the whole assembly cell slows down, waiting for that station to catch up.

Performance can be dramatically improved by just adding simple timers, lights, or visual indicators to help the operators know how they are doing against takt time. Without this external stimulus, operators work without a good feedback mechanism.

Example: Place simple timers and indicator lights at each cell in a multi-station assembly cell. Have each operator push a button when they are done with their work content. When all operations are complete, parts pass to the next station, and the timers reset to zero. Use a yellow light to indicate parts in process, red to indicate over takt time conditions and green to indicate that parts are transferring. An overhead display that shows daily goals against actual production can help operators know if they need to pick up their pace.

9. *Automated part transfers:* Where you have higher volume and consistent part flows automated conveyers and parts transfer mechanisms can make operations more productive. This is a case where high volumes have to drive the economics to justify automated part transfers. When you have the same part movements over and over again, link machines together with automation for controlled flow and lower labor costs. Automated part transfers help keep throughput high while controlling part flow in accordance with takt

time. Labor productivity increases when it is paced by a mechanical method like an assembly line or conveyor.

Example: Take the finger jointing cell example in number five above and make all of the part transfers automatic. With enough volume this investment may be justified by eliminating one or two operators.

10. *Combining machines:* Combine two or three operations into one machine. CNC routers were an answer for this need in woodworking industries. CNC routers can perform work fast, they are designed to do several operations within one machine, and they also have fast setups. However the challenge with large multi-purpose CNCs is that they "want" to be centralized, bottleneck monuments on the shop floor. CNCs are expensive because they have tremendous flexibility and capability at very precise tolerances. They are also versatile, with capability to process many different parts. Since they are both expensive and versatile, we tend to want to keep them working on as many parts as possible. CNCs are normally setup as centralized machining centers, and as such they can create their own batch-and-queue bottlenecks. CNCs are not very successful at one-piece flow in their catalog configurations; however, the concept of versatile computer numerically controlled machines can be adapted to lean flows as in the following examples.

Example 1: If you have a great machine design capability, why not design your own dedicated CNC machines that combine multiple operations into one machine? Interior-door factories need to process parts with many routs for the hinges and strikes. Doors come in many standard heights, 6'6", 7'0", 8'0", etc., and they also come in custom heights. Design a machine that can do all of these routs and end cuts for all heights and machine the left and right stiles at the same time. Think of all the individual process steps and setups you can take out of your factory. This machine is fairly complex with several moving machine heads and will take some advanced process design, but it can be done!

Example 2: A "flying" cut off saw that cuts parts to length as they come off an extruder is another common way to combine machines. The saw moves at the same rate as the extrusion process and makes the length cuts. With this capability, sometimes you can kit parts right off the extruders, ready for final assembly.

LESSON 6:

BEWARE THE MRP MONSTER WITH LEAN SHOP FLOOR FLOWS

Design your lean flows so you don't need shop-floor inventory tracking through materials requirements planning (MRP) data systems.

Today we have more information technology capability than ever. Information flow is a key competitive differentiator and it plays a key role in lean businesses too. However, in the case of lean manufacturing, the old saying that less is more applies to shop floor inventory management checks. Your lean processes should flow so directly and continuously that computerized inventory systems should not be necessary to manage inventory heavily on the shop floor. Visual inventory checks, clearly defined shop floor flows, and defined raw material and work-in-process inventory replenishment strategies should help to intuitively run your lean shop floor.

In well planned lean businesses, you should work to only use computerized inventory to manage the inputs and outputs to the manufacturing process. Said differently, you should only need to manage your raw material and finished goods inventories through your shop floor data system. If everything moves via flow processes and short lead times, using simple inventory management checks, and your eyes, mouth, and brain should be all the shop floor inventory management you need.

In mass production, inventory checkpoints were often woven in throughout the shop floor to manage part movements. In these batch-and-queue systems, there was too much inventory scattered everywhere to simply manage production flow visually. Computerized inventory management systems were installed to help control part flows and provide inventory visibility and to help with financial reporting. However, even in mass production, MRP systems had limited success because of limited real-time accuracy of the data.

If your lean manufacturing processes are designed well with continuous flow, little inventory and short lead time processes, you should be able to see that the process is working with your eyes, not through some computerized inventory management system. Don't slow down lean processes with a host of inventory tracking points. Adding inventory checkpoints adds direct costs to your business to keep the inventory system populated and also adds indirect costs to keep the systems programmed and working. Complex information requirements on the shop floor also get in the way of making continuous improvement changes. The costs and man hours to update these systems becomes an anchor.

Also, it is very difficult to keep complex inventory systems accurate—the more data that is in them, the more room for mistakes. In batch and queue manufacturing, the theory was that MRP would help with shop floor flow. In reality the information was rarely accurate enough to be functional. This is a case where good visual management of defined flow processes is much more valuable than information systems solutions.

Of course there are some exceptions where you may still have to manage WIP inventory through data systems. For example, you may have subassemblies that you build to inventory, to pull from for later demand. Seasonal businesses, like building products, do this to keep their people working in the off-season. Furniture companies do something similar by producing unfinished case goods that they stain and finish later to order. You might still need inventory checks for cases like these, even in your lean manufacturing system. The point is to build your lean shop floor around visual inventory management as the rule and almost never use inventory control and MRP on the lean shop floor.

USE VISUAL PULL PLUS COMPUTERIZED DATA

Information and part flow is also where the concept of *pull* processes comes into play in lean. The concept of pull manufacturing is that your final product demand drives all of the up-stream processes on the shop floor, backward from that final product demand. In other words, if you decide to build product A, product A drives all of the requirements for the parts and subassemblies that go into product A through simple shop floor pull signals, all the way end to end through the plant.

The common lean way to create pull is through visual management systems with *Kanban* indicators, or ticket types of signals, in the plant. A *Kanban* can be a piece of paper (ticket), an empty bin, a cart with labels on it, a display board, or even colored light systems can be used to signal a production need. There are many published resources that you can refer to for specifics on how to create these visual pull signals in the plant.

Often lean experts will push you to try to run your shop floor only off of pull signals without relying on computer data much, if at all. The reason for this direction is that visual pull will force you to simplify your process flow so that human and machine interactions can easily run the process themselves without relying on outside data systems. This is a good general direction with strong potential benefits in process reliability and throughput because the plant controls its own destiny with simple, human interactions and visual flow.

However, in my experience, continuous visual pull is rarely the best way to run the entire production process. Most good lean manufacturing businesses still use a combination of direct visual pull along with order data to run the shop floor effectively. Businesses with highly configurable end products, those making complex configurable components, also benefit greatly from "pushing" data to machines to make these parts. This is where shop floor data helps to enable lean; in fact it can be the best way to do highly configurable component manufacturing.

Think of making kitchen cabinet doors to order. We talked earlier about the large offering that a cabinet business might have. As a simple example, if we have eight wood species with twenty unique profiles in our cabinet door offering, we need to inventory one hundred sixty unique species-profile combinations to have all of these lineals ready for pull in a cabinet door-making grocery store. That usually won't work; it's too big and too expensive to inventory all of these combinations in a big cabinet business that might make several hundred cabinet doors per day. What you might do is inventory a grocery store of the high runners and then manufactures the others to demand. In order to make these, you are going to need to know how many feet you need across a day or two of production, depending on the system you setup. This is a perfect situation where pushing data from your order system to look forward will help you be more effective by producing the profiles to a store for when you need them. Data can also be used to help with cutting down the lineals to final part sizes for better yield. It's pretty tough to be really effective at these types of operation with purely visual pull signals. A combination of pull signals and "push" data is very typical as we look for lean solutions across the shop floor.

USE DATA-DRIVEN PROCESSES

Data, machines, and people can interact to make lean more effective through data-driven processes. As covered above, using MRP systems to manage inventory flow on the shop floor can be an impediment to lean flow; however, using your information system data to help setup machines automatically is a great capability that can enhance lean flows. Send your order information right to machines to have them automatically setup for configurable part making. This item was included in Lesson 5 as a tool to design lean-flow processes.

Data-driven processes provide a powerful lean capability for businesses that have configurable parts. A national art framing company is a good example of the use of this tool. I was exposed to a company that was making thousands of art hangings to order every day. The art was printed off, the wood framing was measured manually and cut to size from prefinished lineals, and the glass and mats were also measured manually and then cut from sheets. Finally all of the parts were assembled, and the final product was packed and shipped.

The challenge in this business was that none of the dimensions or offsets were consistent from one frame or print to the next. The border around each print was unique in each case, and the offsets for each style of wood frames were unique too. As a result the parts preparation was being done by purely manual methods. Machine operators were responsible to figure out these dimensions individually for every product that was ordered. The production pace in this business was much lower than it could have been, and scrap and rework were very high as a result of these manual processes.

Data-driven machines helped this business to be able to cut down frame parts, matting and glass automatically for high volume sellers instead of having operators do it manually. This required that the dimensions and offsets for each high volume print and frame be accurately entered into the business's design information system. This information could then be translated to the shop floor machines through a set of math calculations. All of this is relatively straightforward to do with today's machines and information technology. Also, as this business grew, it made sense to centralize their high volume glass and frame cutting operations to improve productivity, yield, and quality. This business implemented just such a system, starting with their highest volume offerings. Once they implemented these data driven processes, the business also learned to optimize the yield of their frame lineals and glass by looking forward into the day's production. The overall productivity improvements were impressive, and throughput increased more than twofold.

The point is, don't rely only on just your operators to get configurable machine operations right. Businesses with highly configurable product lines operate better by directly driving the data to automated machines and processes.

PART 3

AVOIDING BOUNCE BACK FROM THE LEAN SYSTEM

LESSON 7:

LEAN NEW PRODUCT DEVELOPMENT

> Short lead times are your only lean solution, and you must design them into all of your new products from the start!

Your new product development process is going to need to change for you to create and sustain a lean business. After your initial thrust to clean up the shop floor and get existing products to flow, your new product introductions are going to have to follow these same principles to keep the business lean and take it to shorter lead times.

Here is how new product development works in lean organizations. We've learned earlier in *The Lean Hangover* that lean is about building products cost effectively through direct flow manufacturing in short lead times. If you are successful with lean, these short cycle times are going to become the one and only way you manufacture. With the lean transformation, the process starts with your existing products but soon becomes the case for your new products too. When your product engineers start working on new product initiatives, they need to bring in the Manufacturing Systems Engineering (lessons 2–4) team right from the start to work out any process challenges that might get in the way of lean. New product development and new process development have to be done concurrently to ensure you have great manufacturing flow for every new product. You cannot separate product from process design, they go hand in hand. Every new product must be able to be manufactured with direct flow and with short lead times. The point is short lead times becomes the only formula for running the business as you commit to lean manufacturing. Short lead times are your only solution, and you design them into all of your product flows.

Here is simple model as a way to exemplify this. Lean businesses often design their flow within a few standard manufacturing lead times for their product lines, and something similar to the chart below could apply to just about any light manufacturing business.

Activity	Standard Products	Custom Products
Order Taking	Day 1	Day 1
Manufacture Parts and Sub Assemblies	Day 2-4	Day 2-5
Assemble and Finish	Day 5	Day 6-7
Pack and Ship	Day 6	Day 8

When you develop new products, your lean organization is going to learn to stay consistent with these lead times, always. These manufacturing cycle times become the way we run the business - they become the rules to running the shop. In mass production lead times could be long and vary from product to product. Long, variable lead times were innate to batch and queue manufacturing and we had finished goods inventory to service customers adequately. In lean we build in short, reliable lead times, always!

However, new product initiatives can bring about new processing issues that can challenge your short lean lead times. Your new products are going to offer features and functions that you've never manufactured before. This is what makes them new and innovative. Sometimes that involves introducing new manufacturing processes that aren't capable of short lead times. Parts that need to wait for long cure times or long burn-ins are simple examples. Or new processes that are "monuments" with complex long setup times can challenge your organization to be pulled back into batch processing. The nature of doing something you've never done before can challenge your operations with processes that were not developed for short lead time, cost effective, lean flows.

Once you've built your business around lean, don't let it slip backward with new products that require long processing times. Keep manufacturing engineering involved upfront to develop short lead times for each new product. This is, once again, why you need a Manufacturing Systems Engineering team. You are going to charge this team with finding lean-flow solutions for new products. These are the people that are going to work with the new product designers to develop continuous-flow process solutions. The process will not be straightforward for every new product idea and will require true product and process development. Your process engineers will have to invent solutions to create continuous flows for your manufacturing system. This is the essence of what keeps lean manufacturing functioning as you innovate with new products.

There are two lean business processes that can help you with this challenge of designing new products and new processes concurrently. These are called 2P and 3P kaizen events. 2P and 3P kaizen events are lean tools that can be used to create robust product designs and processes that work together to ensure lean flow.

2P stands for a production preparation design kaizen event, kaizen events where product and process engineers work together over a focused time frame to develop the conceptual design of *strictly process flows*.

3P stands for a product and production preparation design kaizen event. 3Ps start with a clean sheet product development slate to create and test potential product and process designs that require less time, material, and capital resources. The purpose of 3Ps is to generate robust, capable, lean process designs out of new product concepts. 3Ps are kaizen events where product and process engineers work together to optimize product and process designs concurrently so that the new product can be manufactured using lean processes.

In summary, 2Ps are focused on nailing down the process flow once the product design direction has been set while 3Ps are focused on setting robust product and process design direction while the product design can still be changed. Both of these are excellent lean new product development tools where we gather the product engineers, process engineers, and marketing people together to flush out design ideas and turn them into stronger product and process designs. Following are the details of each of these kaizen tools.

2P KAIZEN EVENT

A production preparation (2P) kaizen event is made up of engineers working together for a focused number of days, normally three to five, to work out the basic concepts of process flow for an existing product design. These events use brainstorming and "trystorming" activities to develop the process design. The goal of these events is to eliminate batches by developing continuous-flow processing solutions. The 2P team starts off brainstorming ways to process parts and assemblies that can create lean flows. The group uses fishbone diagrams to show how parts and assemblies move from step to step in the process. The main spine of the fishbone is populated with each step in the process, and the branches are labeled with the actions that occur at each step, like "bend," "punch," "shear," and "coat." The team identifies current process technologies that can accomplish these actions. Teams use 2P events to brainstorm small, inexpensive machines to replace large, expensive equipment, eliminating these large monuments to improve lead times, quality, and productivity. Another tool is to leverage examples in nature that do these same actions. Coat might be shown by the way an elephant sprays water on its back to stay cool. This is a way to get the creative juices flowing and expose the team to many different conceptual solutions.

Then the team breaks up into small groups to hand sketch how they would accomplish the process in question. This is where the team uses the concepts I covered in Lesson 5 to create flow in their solutions. At the end of the brainstorming portion, the small teams meet to present their ideas and vote on the best concepts. This is a way to flush out the two or three best alternatives and take these mainstream process designs into trystorming.

In trystorming the kaizen team uses cardboard, wood, foam core, PVC pipe, and other readily accessible materials to mock up the primary process designs. Trystorming is a great tool to visualize the actual process flow and identify issues and complexities within it. At the end of a 2P event, the team chooses one or two process design concepts to take into actual development. Assignments are made to start developing the specific machines and business cases. A new product development program may use several 2P events to finalize the process design. These 2P events are very useful for nailing down process designs.

3P KAIZEN EVENT

Where 2P events assume the product design is fixed, product and production preparation (3P) events focus on defining *product design directions* so they can meet customer requirements with efficient, capable lean-process flow. The 3Ps start with a product development concept to create and test potential product and process designs against, determining designs which require less processing time, material, and capital resources. Product and production preparation kaizen events focus on concurrent product and process design to ensure the following:

1. Newly identified customer needs are met or exceeded by the product design
2. Product designs are optimized for cost and manufacturability
3. Product design are supported by capable, efficient, lean process flows

3P events gather product engineers, process engineers, and product marketing people together, typically over a one-week time frame, to identify and evaluate alternative ways to meet the customers' needs through robust product and process designs. 3P events are used to develop products that are lower cost and easier to manufacture. Product and production preparation kaizen events work to connect process steps in new continuous flow manufacturing through custom designed, right-sized equipment. They represent a dramatic shift from the continuous, incremental improvements of existing processes through CI kaizen events. Instead, 3P events seek to develop "quantum leap" design improvements that can improve manufacturing performance to a level beyond what can be achieved through the incremental improvement of existing processes.

The steps of a typical 3P Kaizen follow:

1. Review product design concept.

2. Brainstorm and diagram initial manufacturing process flow.//

3. Identify manufacturing challenges to low cost, continuous flow production.

4. Brainstorm solutions to these challenges including both product and process changes.

5. Brainstorm opportunities for combining processes into linear manufacturing.

6. Identify opportunities for custom designed processes and machines.

7. Mock up alternatives for complete production process.

8. Evaluate alternatives for cost, quality reliability.

9. Choose primary design alternative for development.

Some of the same processes covered in the explanation of 2Ps are also used in 3P events. Fishbone diagrams, brainstorming, and trystorming are used in the exact same way in 3Ps for new product programs to narrow down the process design alternatives. In addition, the following tools can be useful in a 3P kaizen event.

Design for manufacturability (DFM) is a structured tool that is used in 3P events. There are many available resources and literature on the details of design for manufacturability, but in brief DFM is the philosophy of value engineering product designs to reduce parts count and labor content, making product designs more cost effective. Design for manufacturability is a methodology that measures part count and work content in an effort to reduce those as you value engineer new product designs. As an example, brute force engineering uses a screw and a nut to hold everything together while elegant design uses snap together pieces to hold parts together. Another simple DFM example is integrating a metal bracket into a plastic molded part to combine the two parts into one. Significant improvements can be made when we sit a group of engineers down with the intent to simplify product designs. This is a great process that should be applied to every new product initiative.

Competitive benchmarking and *target costing* should be done as an ongoing effort in your business. Dedicate a competitive benchmarking area in your company and populate it with relevant competitive products. Analyze them for the data that is important to your business. Part count, labor content, pricing, and bill of material cost estimates are all good data points to track. This is a great tool to use with new product developments as a check of your product's competitiveness. Hold your product teams accountable on how they are going to beat these competitive designs with their lean new product introductions.

For the sake of completeness, a more difficult to use but sometimes highly regarded 3P tool is the *quality function deployment (QFD)* matrix. Quality function deployments matrices were developed to use in the earliest stages of new product design to evaluate voice of the customer wants and needs against known product and process design alternatives. Quality function deployments have often been promoted as *the* lean tool to translate customer wants and needs into actual product designs. Businesses obviously want to invest in new product technologies that meet specific defined customer needs and wants. The intended purpose of a QFD matrix is to ensure customer needs are translated into real product designs—cost effective designs that can manufactured with reliable, lean processes.

As an upfront bias before I explain this lean tool, QFD's have had marginal real effectiveness in my experience, as I will explain below. I include QFDs in this discussion to give you exposure to the tool and because of their strong association by many with lean new product design activities. Maybe you can figure out how to use them better than I have.

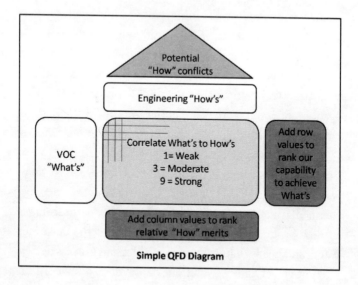

Another name you may have heard for a QFD is the *house of quality,* or *Toyota house of quality.* The matrix looks like a house, therefore the name.

In its simplest form, the voice of the customer needs are first identified and prioritized on the left. These are the "whats." Next, potential product and process technology alternatives are listed across the top. These are the "hows." Then the whats are correlated with the capability of the hows to accomplish them by a numerical ranking. The VOC priorities are added horizontally, and the technical correlations are added vertically to give relative rankings.

Strong scores across the horizontal bar on the bottom indicates that we have important potential technologies that could meet new customer needs to stay ahead of the competition. Conversely, weak scores here can identify potential new technology gaps that should be researched and developed for use in future products. These weak technical capabilities may be able to provide potential competitive advantage to achieve the voice of the customer whats and should be considered in your research and development portfolio.

Strong product performance ranking on the vertical bar on the right indicate that the business has several potential alternatives to achieve this particular customer requirement. We have a stronger confidence that these are doable, and we should target them for the final product specification.

The roof of the house shows support or conflict amongst the product and process technologies. One potential technology might support or detract from another, so this is a place to identify that as we choose the technologies we will pursue. Frankly, I've never really figured out how to use the roof to change my decisions.

Quality function deployments are not a perfect quantitative tool; you have to check the results against your experience and intuition. Quality function deployments are meant to flush out the relative capability of our existing knowledge base to meet articulated customer needs. Think of QFDs as a high-level way to make tradeoff decisions on customer needs versus primarily *existing* product and process technologies. A QFD is probably not going to be the place you are going to uncover the untapped, unarticulated product breakthroughs for your business.

The theory of the QFD case is great, but it's been difficult in practice to QFD matrices effectively. I've frankly found QFD exercises to be better time to catch up on your sleep than to design great products. They become quite complicated and unwieldy as we list all of the potential whats and hows. Another potential shortcoming of QFDs is that the results are driven only off of the priorities and the correlations we plug in subjectively. The numbers are based upon gut feeling and group-think many times; change the subjective correlations for a couple of technologies, and you will get significantly different results. Maybe I'm missing something with QFDs and you can figure out how to use them effectively.

While we are on the topic developing great new lean products, there is one other important product development pitfall that needs to be addressed as you develop your lean new product development processes. How many times have you purchased something brand new and when you first started to use it noticed something really out of whack—one of those "What were they thinking here?" moments? Or have you been in business and watched a great new product being developed with all of the hard work and excitement leading up to what the business thought would be a seamless product launch only to find some major performance issues once the product is in the field? Have you ever wondered whether the people who designed these products have ever really used them?

Chances are they haven't, or at least, not enough. We all get caught up in the "corporatetocracy" of doing our jobs, especially mangers and senior-level leaders, and sometimes fail to spend enough time actually trying out our new products. Here is how this plays out. The division leader is busy, busy, busy. The engineers are working on the new product, and they keep Mr. Division Leader informed about how great everything is going: "Best new product design ever; we are all feeling great about it," "The lab is doing all the extensive testing to prove out the new designs, and they should find any potential problems," "We are feeling really good about this launch and should be ready to go in record time." Leaders can be lulled into a sense of comfort by the process and let the team push products to market too quickly, before we do significant human interaction testing.

Lean businesses need to launch products that are well thought out and ready for the marketplace. To help guarantee successful new product launches, you need to *play with your new products*. Unproven new product designs create waste through lost sales, scrap, rework, redesign, and a host of other headaches. You need to setup processes to actually use your new products, to make sure they are well thought out and ready to launch. You need to have internal and external clinics and use extensive field-testing to prove out your new product designs. Put them in your house, your car, your lunch box, or whatever applies. Drive, ride, push, and pull your new products. Find the right formula for your business.

This sounds rather intuitive, but you would be surprised how often businesses fall out of doing this simple practice. We put the cart before the horse and rush our products to the market place before doing the necessary due diligence with field testing and human interactions. I've seen this become more prevalent over the past ten or so years as many lean business have chased shorter new product development lead times to a fault. They try to translate lean's short manufacturing lead times into shorter new product development cycles without building the robust processes and back wheel R&D that it takes to do this effectively. Lean businesses like Toyota have gained significant publicity for their effective, short new product development timing. That's great, but it comes from a better system of R&D and new product development steps including significant prototyping and field testing. Too many businesses have omitted deep, real people interaction as an integral part of their new product development as a way to cut their new product development cycle times.

This activity is a key to launching new products that customers will love and ones that won't cause a lot of waste in your operations through field quality issues, rework, and redesign. Don't get burned by this short cut.

And, the boss needs to play with his new products too!

LESSON 8:

BUILD-TO-ORDER IS NOT THE GOAL OF ALL LEAN!

Many lean businesses will still never effectively achieve build-to-order manufacturing because of their fundamental business models

A good lean program results in shorter lead times for your product flows. We've talked about this as one of the fundamental lean principles. Many businesses are able to leverage this short lead-time capability to their advantage to begin to *build to order (BTO)* and/or *assemble to order (ATO)*. The theory of the case is that when you can build your products in a few days because you are lean, why not wait until the order is in hand and then make what you need *just in time (JIT)* to fill that actual order. This is why the name "Just-in-Time" manufacturing has become synonymous with lean manufacturing.

Creating a true build-to-order capability is somewhat like having a goose that lays golden eggs. It is one of the best ways to reduce inventory levels, improve customer responsiveness, and increase return on investment for a business. You build only what you need exactly when you need to. And continuous-flow manufacturing enables some businesses to achieve build to order through driving manufacturing lead times down to just a few days.

However, many lean practitioners confuse that build-to-order is the objective of all lean manufacturers. Build-to-order is not the end goal of every lean program. Many businesses will never be able to effectively use build to order due to the makeup of their business model. The furniture business in Lesson 12 is an excellent example of a business model that absolutely should not be chasing build-to-order for reasons explained in that lesson.

Let's back up and explain what build to order and assemble to order (ATO) actually are. Build to order and ATO are very close to each other. The major distinction is that in assemble to order you simply assemble stocked components to order to create your finished goods. By contrast, build to order also entails fabricating the upstream components to order also. An example of ATO is Dell Computer's manufacturing model. They assemble modular computer components to fill your custom computer order. They are not fabricating the motherboards and the hard drives to your order, just assembling them into a computer. An example of BTO is a kitchen cabinet factory that machines your doors, drawers, cabinet boxes, and other parts specifically for your order and assembles them to order.

For simplicity, and since they are so close in practice, we will use the term "build to order" to include both cases, BTO and ATO.

Build-to-order capability makes for a very efficient business model, with lower inventory levels, improved capital leverage, and lower overall costs. A business that builds to order ensures that it only builds what is actually ordered, so it has money tied up in only what it needs to successfully fill orders. Inventory turns are very high in these businesses. Stock outs and order fill delays also become rare, and excess and obsolete costs improve because these businesses have little inventory on hand that could actually become excess or obsolete.

On the other side of the spectrum are *build-to-inventory (BTI)* business models, where we build products to inventory to fill orders from that inventory. In build-to-inventory businesses, we manufacture items and place them into finished goods inventory to sell from that finished goods stock. In BTI businesses, production schedules are normally set based upon sales trends and sales forecasts. Current sales rates are evaluated against inventory on hand and production lead times to determine when we need additional production volume. Also, production is planned with some level of finished goods safety stock to cover unpredictable changes in sales rates and delays in production schedules.

Build-to-inventory businesses are innately challenged to predict sales patterns and trends so that production can stay ahead of these sales forecasts. Because sales demand is always unpredictable, by nature BTI businesses never get their production forecasts perfectly right. Safety stock inventory is normally used to help ensure high fill rates over a long time frame as demand swings up and down.

However, this safety stock is never able to cover all situations, so when sales unexpectedly surge, or production gets behind schedule, stock outs become a problem in build-to-order businesses. Also, because these BTI businesses are building to forecast, they often end up with higher excess and obsolete costs due to the unpredictability of demand.

Build to order is a hands down, innately more efficient business model than is build to inventory. This is why lean manufacturing businesses in general drive toward build to order as the preferred state. But here is where the confusion can start to come in. Many business leaders misunderstand that BTO is the desired end game for all lean businesses. Since lean, at the core, is about driving short lead times through continuous flow production, it is a natural carry on to think that we are trying to eliminate all build-to-inventory activities through lean's very short lead times. Lean consultants and other lean teachers often push their clients to drive toward build to order as the ultimate goal of their lean efforts. As a general direction we are driving for lead times that are as short as possible, and for many businesses, BTO is the end game.

However, in many lean businesses BTO still cannot be practically achieved. Build to order is not going to work in every business due to conditions that are built in to certain business models. Following are the four conditions that I've experienced that result in the need to continue to build to inventory:

1. Long processing lead times, even after lean

2. Potentially low, unpredictable product volumes with high product turnover.
3. Large seasonal variation
4. Long lead time imported goods

Let's cover each of these conditions in some detail.

1. *Long processing lead times, even after lean:* Some businesses are still like art. Laying up highly figured veneer for a furniture dining table top still takes a long time even in a lean business, primarily because this work is done by hand. A complex veneer face may contain ten species of laser-cut, hand-fitted, wood pieces. Once completed, this veneer face is pressed onto a wood substrate to form a veneered panel that is ready for machining into the desired final shape and dimensions. The machining process can involve several machine operations and take several days. The overall process of making complex veneered panels takes very long, but that is just an unfortunate fact that goes along with the fine furniture business model.

    If you still have processes with long lead times even after your best lean efforts, you have one of two choices: build to inventory or convince your customers they should wait a long time while you build their products. Build to inventory usually wins.

2. *Potentially low, unpredictable product volumes, with high product turnover:* New product development in the furniture industry is like making spaghetti. You know

the old saying about how to determine if spaghetti noodles are cooked well enough—you throw them at the wall, and if the noodles stick, they are ready! The furniture industry cranks out a tremendous number of new products every six months, and to see if they have it right, they throw the designs to the market place to see if something sells. The industry introduces 10 to 20 percent of its product line, brand new, every six months, at the spring and fall High Point, North Carolina, furniture markets. The furniture industry is on a very hectic new product development cycle, but that is a story for another time. The end result is you never know which of these products are going to sell well. Also, there isn't enough time to do any significant product engineering, process engineering, or market research on these new product concepts. The designers throw their best ideas at the factories, the factories build and refine a couple of prototypes, and then the business throws the new products to the market! And furniture businesses do this over and over again every six months. So, when a product sticks and sells really well, you've got to run hard in the factories to keep up with demand. However, in reality, most items sell just a few hands full of pieces, and then they are removed from the line.

There is no cost- or time-effective way to setup continuous flow, build-to-order manufacturing process for all of these complex, short-lived, unpredictable new products. With the furniture industry cranking out new products every six months while purging their poor selling ones at the same time, it's nearly impossible to justify the cost to setup continuous flow production for all or even many of these products. And the necessary time isn't available with this rapid cycle of new product development. Think of it this way: the investment of engineering effort per piece sold would be huge to invent ways to flow these products through the factory in dedicated one-piece flow cells to create build to order. In retrospect I've been in furniture factories that tried to do exactly this for a few of their high volume models to show how lean could work, but what these businesses found was that a one-piece flow, BTO strategy only worked for a handful of products while the remainder had to be processed through traditional mass production methods.

Instead, in furniture businesses you need to setup bulk lean cells for common part flows across the product line and then handle all of the uncommon flows with traditional routings. The non-lean routings go into a spaghetti bowl of machines where each part takes its own path through up to ten or fifteen machining steps. Each of these machines still has unique setups for many of the part designs, and, inevitably the setups are still fairly long, even after lean setup reductions in the furniture factories I've worked in. There is just too much setup cost to run small batches or one-piece flow in a business model like this. The products would cost way too much if we were to run small batches. Nonetheless, some furniture businesses have become confused about their own business model and tried to do just that, chasing build-to-order too early through their lean programs much to their own dismay!

Businesses that still have long machine setup times, even after their best lean efforts, are still going to need to use batch manufacturing to have any chance of being cost effective. Further, couple batch processing with long manufacturing lead times and these furniture businesses are forced even today to build to inventory as their only alternative. Taking this one step further yet, the unpredictable sales and short product lives in the furniture business means their build to inventory model is a difficult one at best. Stock outs are common for items that have high sales rates because it takes too long for production to catch up with demand. Excess and obsolete costs are high because production builds large batches to sales forecasts that often miss the mark significantly.

This business model isn't what we would call a great lean model; in fact it's probably as far from lean as is possible. But it is what it is, and that is what makes it important to understand the fundamental shortcomings of our manufacturing systems to plan for the right lean approaches. What this means is that furniture businesses and others like them continue to run a risky build-to-inventory production strategy, one that creates innately higher manufacturing costs.

However, there are still tremendous opportunities for lean in this business model. We need to work on driving the setup times down through better machine designs and common part designs. We can also setup machines in flow cells for common parts routings, as mentioned above, and perform coordinated changeovers as a successful lean method. When we've worked our setup times down to a few minutes, instead of the hour plus it has traditionally taken in the furniture business, only then we can turn on smaller batch sizes and flow parts through the factory much quicker. That has been the challenge in the furniture business for many, many years, and when coupled with the pace of diverse product flows and rapid new product development in these businesses, it has been a very hard nut to crack.

Lesson 12 is an excellent case history about this particular challenge.

3. *Seasonality*: Many buildings products companies sell two to two and a half times the volume in the summer months compared to the winter months. Sales of windows, doors, cabinets, flooring, and other building products surge this way in the summer and fall months. If we run strictly a build-to-order model in these businesses, it becomes difficult to have the capacity and labor to ramp up to a build rate of two to two and a half times during the building season and then take production levels down equivalently in the winter months. Therefore, many building products companies build to inventory in the winter months to help level load their factories. Businesses with significant seasonality often find that building to inventory to level load the business is an effective strategy to get through the slow periods.

4. *Long lead time imported goods:* It takes ten to fourteen days to ship product from Asia or Europe to the US. If you import finished goods or components from these or

other continents, you are going to need to import them to inventory. That's just the way it is.

If you import components you might still successfully run an assemble-to-order-business model within your operations. However, you need to guarantee the quality and reliability of these imported components, or a misstep can shut down your build to order quickly, causing order fulfillment delays immediately.

If you import finished goods, you are going to have to stock them to inventory with enough stock to cover demand variation and the long lead times to receive shipments from overseas. Welcome to the world of imported, low cost finished goods!

Overall, this business model can take a long time to bring new products to market, but low costs make overseas manufacturing attractive to some. Long-order fulfillment times from overseas suppliers normally means your will have to run a build-to-inventory manufacturing program.

Build to order is valuable asset for the lean businesses that can achieve it. However, build to order is not possible for every lean business. Some lean initiatives will result in very short lead times, short enough to enable those businesses to implement a build-to-order model; others will not be able to compress lead times far enough to achieve BTO. Build-to-order businesses and build-to-inventory businesses still both need to be lean where they can. Both businesses need to find ways to create continuous flow to generate the shortest practical lead times.

LESSON 9:

LEAN CULTURES DRIVE EXCESS CAPACITY

A loosely managed lean culture can cause you to over invest in capacity, putting the business at risk when a downturn hits

During periods of strong economic growth, a lean business can over invest in facilities and equipment in an effort to maintain short lead times across a full range of product offerings. Lean cultures can cause unusually high pressure to install additional manufacturing capacity so that all products can maintain short lead times always! Once products have developed into diverse, dedicated shop floor flows through lean, businesses create a huge risk of proliferation of equipment and facilities, much more so than in mass manufacturing systems. The end result of this lean bounce-back risk is that lean businesses can be left holding the bag with painful excess capacity and when the market stalls, too much money tied up in equipment, floor space, and people. This risk is particularly high in diverse, build-to-order, seasonal businesses as explained further in this lesson. Any business that opens this matchbook is going to be playing with fire when the market makes a meaningful correction and revenues start to rapidly shrink. Instead, work to create a constrained capacity model with shared product flows to insulate the business from this risk of over investment.

Following are the details of how this potential pitfall happens uniquely in lean businesses.

This risk has two related factors that cause it to happen more strongly in lean businesses. The first is the growth drug I've mentioned before. During a period of general economic growth, everything charges upward in the business, "a rising tide lifts all ships" as the saying goes. When business is booming, your business units (BUs) are going to add capacity continuously to meet their corporate growth commitments. Business units are rewarded for hitting their growth targets. They have to be ready with equipment, facilities, and people to meet these goals. Not having enough capacity means we don't live up to our new brand promise; reliable delivery and short lead times created through the lean transformation.

This growth pressure is not unique to lean businesses; it exists this same way in mass manufacturing systems. What is unique, however, to lean manufacturing systems is the combination of this universal growth pressure with a second factor that is particular to lean manufacturing systems.

This second factor is that lean manufacturers end up by the nature of lean with more machines dedicated to individual product flows. One of the primary ways we create continuous, lean manufacturing flow is through dedicated machines arranged in continuous-flow cells. We covered this change in detail in Lessons 2 through 4. In lean, we purposefully decide to have separate, dedicated assembly lines for each of our products, oftentimes with dedicated sub assembly and feeder processes for those assembly lines. And, as lean manufacturing grabs hold and matures, businesses design more and more of their flow processes so that they are dedicated to product lines or product families. This is the natural thing to do when the business is growing and we have gained enough volume to justify each of these separate flows. It is very lean indeed, creating rapid, manageable, dedicated flows for each product so that each business unit can control and plan their business around these. In lean, we work to service all products with short lead times through dedicated assembly flows!

It is also intuitive and efficient for lean businesses to align their human resources this same way. Lean businesses often align their factories and product lines in segregated business units to better organize decision making. The business establishes business units that align people to specific product lines with dedicated manufacturing operations. General management, product and process engineering, and plant operations are often aligned within these business units.

This lean model can directly lead to more diversity and proliferation of the product lines. In a sort of chicken-or-egg argument, under this business unit structure the teams have a penchant to diversify their product designs to capture the potential growth opportunities that they believe are uniquely aligned with their specific products. These business units start to make separate, disparate product design decisions to deliver their productivity and growth goals. We encourage this thinking in lean environments through separate business units with separate product flows in dedicated factories.

Sharing process flows and plants and equipment is not the easy, intuitive answer when we have already established separate product lines with separate leadership through these separate business units. As you grow with your lean manufacturing system, you can start to find that less and less of your equipment is shared across product families, especially during a period of general economic growth where each business unit is heavily rewarded and motivated to deliver on their individual growth plans.

Now, let's connect these lean business unit management structures back with the growth drug. During times of economic growth, your business plan chugs along every year with growth targets somewhere between 7 and 12 percent. And your business units request capital to add the necessary capacity to deliver their growth plans. Sure, manufacturing is going to use overtime and productivity improvements to bridge short-term gaps and extend capacity investments as long as possible, but sooner or later the additional capacity has to be added. Some of these capacity investments can be made in small incremental chunks, a few machines here and a few machines there. But other capacity investments must eventually come in bug chunks; an automated paint line here, an assembly line there and a new building over there! Bricks and mortar expansions almost always come in big chunks, and they are expensive. When lean businesses do this, add capacity across each separate, disparate business unit, with each of those having their own dedicated manufacturing operations, bricks and mortar expansions start to multiply across the corporation. Fewer and fewer manufacturing flows are shared, and growth starts to push the existing systems to add floor space in a multiplicative way.

And there are two additional business factors that can further exacerbate this risk of product diversification and over-capacity expansion. These are: (1) businesses that are complex build-to-order businesses, and (2) businesses that have high seasonal variation. Building products industries like windows, cabinets, and doors fall into both of these buckets. They offer so many product sizes and variations that these businesses are virtually forced to build every product to order; there is no way to build to inventory effectively with the multitude of product shapes, sizes, and feature offerings. And they service the seasonal building industry which has huge increases in demand during the summer months compared to wintertime.

Businesses with high seasonal variation require huge production capacity levels to handle peak demands. Again, many building products companies have seasonality where, from wintertime to the heavy summer building season, volumes increase by two to two and half times. Build-to-inventory businesses could use the slower months to build inventory and smooth out the peaks. Conversely, build-to-order businesses must continue to build to order in short lead times during both the slow and the peak seasons. So when we couple significant seasonality with build-to-order businesses, we create a business model that needs an even greater amount of manufacturing capacity to cover peak demands.

Okay, so let's list out these factors that can cause successful lean businesses to invest in excess capacity:

- We have a build-to-order business.
- We have a trend of actual growth with continuing annual growth targets.
- We have significant seasonality or demand variation.
- We've grown into separate business units, each managing dedicated product designs and manufacturing flows
- We've become proficient with lean and work to service every order and every product with short lead times.

Sound familiar to any of you lean leaders? I'm betting this list does. I've worked in a couple of lean businesses that ended up in this very situation. These are a combined set of circumstances that often come directly out of being great with lean! And under these circumstances, lean business leaders can be lured into supporting all of the capacity investments that the business units believe they need to cover each year's growth projections. We need capacity everywhere to service every product with short lead times every day. This situation can add up to heavy *CAPEX* spending and a ton of risk, depending on when the major investments happen and how these decisions align with an eventual market pullback.

Put these factors together, and lean can convince a business to march right off this excess-capacity plank during strong periods of growth. When the growth finally stalls, your lean business model has you holding the bag with way more capacity than needed. The potential ramifications are strong; layoffs, plant closures, asset write offs, and business restructurings are not unusual.

This dynamic is one of the toughest to avoid in lean businesses. Your business units actually work hard, in a way, to create this problem. They are motivated to ensure capacity always covers demand, and with some level of safety margin. In truly lean businesses when demand outstrips capacity, order fill rates drop quickly, and we start to miss customer commitments almost immediately. There is no inventory as a buffer to maintain delivery commitments. Managing this dynamic takes a really strong quarterback in the CEO or COO role, one who can make the tough calls on prioritizing investments to manage capital investment spending against forecasted capacity needs.

If you really want to be successful through the market ups and downs, don't over capacitize; figure out ways to run a constrained capacity operations model. If you walk through your plants and your new investments are way underutilized, you are probably walking down this plank! Pray that the market keeps growing so you can build your sales to pay for all this equipment and floor space.

Or, better yet, get ahead of this game and protect your company by making sure you can sell most of your capacity, most of the year, even during growth periods. That is the safe way to always be prepared for a market downturn.

Here are several strategies you can use in a lean business to avoid this risk:

1. Share common platform designs so you can share assembly and parts making equipment. I've been in businesses where the good, better, best platforms grew into totally separate designs to serve the same basic functions with really no good justification to allow them to take on separate lives on the shop floor. Leverage common platform designs so that you can share more of your operational capacity across platforms.

2. Force your capital-intensive sub-processes to be "product agnostic." Share paint lines, for example. Design flexible sub assembly lines that can make chassis or components for two or three separate products.

3. Manage your operations and engineering separate from your business units so they drive efficiencies that are important to operations. Build similar products in shared factories. Force product designs to have common manufacturing flows through strong central engineering and operations linkages.

4. Decide not to have enough capacity for everything every year. Delay order lead times when you hit capacity on certain product lines. This may eventually turn away business due to long lead times, normally a lean no-no, but this can be an effective trade-off to stretch a year or two out of existing capacity. If you have the ability, best to do this on your low-margin products to protect profits.

5. Push your existing plants to 120 to 140 percent of capacity and delay major bricks and mortar investments for one or two seasons until you are confident you

can sell greater than 50 percent of a new plant investment. Use overtime and increase staffing to get through these times.

Over the long haul, you will protect the business and make more money for your investors if you don't allow your lean culture to over capacitize!

LESSON 10:

LEAN AND CI CAN KILL INNOVATION!

Lean businesses can stumble by becoming so focused on continuous improvement and running great operations that they consume their R&D capability to innovate.

Lean and the *continuous improvement* (CI) culture can produce strong improvements across all aspects of the business. Kaizen events are used by lean companies to identify and drive productivity regularly on the shop floor. Your engineers will use 2Ps to 3Ps to plan out significant new product and process improvements, designing machines, jigs, and fixtures to improve new and existing processes. Quality improvement kaizen events will be used to redesign products or processes to improve quality problems. And *business process kaizen* (BPKs) events are used to streamline administrative processes, like order entry or on boarding new employees.

If you take this lean journey to heart, making it a strategic initiative, lean and kaizen will grow into the tools of choice to drive improvement across your business. With your team engaged, you will start to see multiple kaizen events happening each and every week. Plant managers will use kaizen events to generate new ideas take cost out of their operations and to make them more efficient as volumes and business conditions change. The lean toolbox will provide the processes for your engineers to make new product launches better. Your finance staff will use BPKs to streamline how they do the monthly closings. The list goes on and on.

In lean businesses we build a *continuous improvement culture*, which works to improve the business by producing profitable results. And continuous improvement kaizen events become *the* way you make the business better. Continuous improvement becomes a core capability in the business and CI improvement targets end up on everyone's annual objectives.

Great, because we need this CI spirit to compete! This is the right direction for the business. We use our continuous improvement culture to make the business better every day, and kaizen events are the way the business drives cost out.

But there is also a slippery slope here; there is a difficult pitfall that develops in successful lean organizations. Continuous improvement and the kaizen culture can become too strong. Continuous improvement can become an overly dominating fix-all to run the business such that it consumes your engineering resources, including the creative ones that should be working on the future innovations and longer term product reengineering for the business.

There are a couple of specific reasons this happens in a lean culture. First of all, CI becomes so effective that we start to use it to fix and improve everything that needs to be made better. Every time the business has a challenge, we use kaizen to fix the gap. A machine breaks down; have a kaizen event to study and fix what went wrong. A particular product has manufacturing challenges; have a kaizen event to improve the design. A supplier has a quality problem; have a kaizen event. We need to take cost out of a product; have a kaizen event.

As was mentioned earlier, kaizen events become the way that we reduce costs and take waste out of a lean business. And, as we become proficient, the volume of these events increases substantially. This, in turn, starts to place a higher demand on our engineering resources, more so than any other department in the business since engineers are involved in making all the product and process changes.

The second part of this dynamic is that CI becomes "the way" we generate productivity to meet the annual cost reduction targets for the business. Great businesses drive productivity targets as a part of managing their annual operating costs. These productivity expectations stimulate continuous improvement projects that almost always need support from the business's product and process engineers. The engineers have to be involved to make these cost reductions. Changes on the shop floor need support from operations and product and process engineering. Changes in purchased materials to lower costs also require engineers to design and validate new components. As a result, our engineers get tied up heavily in the need to deliver this year's plan, and kaizen events are the way they do that in lean businesses.

So we can start to create a bit of a continuous improvement monster here. The annual productivity needs of the business coupled with an aggressive CI culture starts to overwhelmingly pull for more and more engineering resources, including the corporate ones that should be working on future innovations and product direction for the business. Continuous improvement helps us deliver this year's plan, but who is working on the next material breakthrough, the next process breakthrough, the next green technology, the next energy improvement?

One can argue that this is not a "lean" or "CI" dynamic alone. That is true; the pressure to deliver annual plans exists in every business, lean or mass production. And the pressure to use corporate R&D resources to fill gaps in near-term needs exists in all businesses. But there has always been a tougher struggle to keep divisional engineers "division focused" and corporate engineers "corporate focused" in lean businesses.

Here is why this happens. In traditional batch-and-queue businesses, a strong CI program usually doesn't exist, or at least not nearly at the level that it does in a lean business. Continuous improvement is a lean thing. In a traditional business, the plant engineers run the plant and operational productivity projects while the corporate engineers do the project work for major new product or process programs. The corporate engineers do work that is planned and decided to be healthy for the long term on the business by corporate leadership not operations leadership. In a traditional mass production model, there is more of a dividing line between the corporate and the factory engineers. Corporate engineers are not pulled into these types of CI events on a continual basis.

In lean businesses, the line between engineering and operations becomes very fuzzy. We create a much stronger day-to-day pull for our engineers to be part of the CI processes that make *operations* more successful. Any significant kaizen event that affects product or process design requires engineering involvement. A dynamic develops where more CI means more Corporate Engineering support. Over time, operations gain more access to the corporate engineers to support the major product and process initiatives that they need to deliver the business plan. And pretty soon, operations want accesses to decide what all of these corporate engineers are working on too (see sidebar). This gradually becomes a "normal" part of the CI momentum and the way we do business.

Don't misunderstand the point; heavy engineering involvement is not a "bad" lean thing. It is a necessary and desired part of shop floor CI. But because CI is so powerful and because we engage more engineers as a regular part of these activities, the operations side of the business pulls harder and harder to consume more engineering resources, to deliver their near term annual operating plans.

You've heard about lean killing innovation, businesses that stumble because they become so focused on running great operations that they lose their capability to innovate. That is exactly the point. This lesson is really how the cultural momentum of lean and continuous improvement can swing the pendulum strongly to the side of "fix stuff right now" and forget to leave some resources behind to work on the future products and processes that the business needs three to five years out.

Of course we all realize that CI can't fix everything. We also know that lean cannot develop the product breakthroughs and innovation that will take the business into the future. But even with this knowledge, it's really common for this lean organizational dynamic to just happen. Left unto itself, a strong CI capability and the annual pressures to deliver this year's results will cause this pitfall, like a whirlpool sucking everything into its vortex. This is a subtle dynamic and one that many senior business leaders mistakenly fly too high over.

In lean businesses, we need to be particularly committed to create mechanisms to protect the R&D engineers who reengineer our product lines, or we will be stuck in CI rut, continuously CI'ing the products and processes we currently offer. You have to protect these resources in a lean business, or the pressure of CI will envelope them. Left between the divisional leaders and the engineering leaders of the business, this becomes a never-ending dogfight, and the near-term CI needs usually win out.

The best way to mitigate this in my experience is to decide at a corporate level how much Corporate Engineering/R&D you really want to do, then staff it, fund it outside of the divisional P&Ls, and put a fortress around it. Otherwise CI and then near-term business performance will consume it!

Another method that can help manage this conflict is using the concept of *sustaining engineering*. Sustaining engineering is the practice of consciously deciding to leave a mature product design alone to live out the next few years of its life basically as it exists today. We assign a minimum number of engineers to a product that is being sustained, to support just the critical issues. When you drive productivity and CI improvement expectations across all of your product lines equally every year—like spreading peanut butter on bread—you will get people working forever, spinning their wheels to find ways to improve all products, including mature designs. However, improvement opportunities taper off as products mature. The right answer is to pull back on this productivity pressure for your mature products, ones that are three to five years or more into their product life cycles.

We put products into the sustaining engineering mode so that we can free up technical resources to better balance the engineering resource plan and allow engineers to work on new product designs that can leapfrog existing ones. It takes strong leadership and discipline to just "sustain" some products and accept their current performance levels for the time being so that the business can create new, better products. But you need to have deliberate plans to have resources work on the greater good. You need this sustaining engineering tool in particular in a lean culture because of all the pressure and CI expertise running hard to deliver productivity on every product, every year.

It takes leadership direction, discipline, and good product portfolio management, usually at the corporate staff level, to be successful with keeping resources dedicated to R&D within a lean business.

LEAN ORGANIZATIONAL DYNAMICS AT WORK

Here is a typical sequence of events that shows how lean creates this particular pitfall.

Put on your Division President or general manager hat for a moment. I'll give you some time. Okay, now you are ready. It is time for the annual planning cycle in your business again. Hooray! Let's skip most of the fun work, the long hours sitting in meetings listening to the marketing people telling everyone about all the important changes in the marketplace and how we are perfectly positioned to take advantage of them, and let's cut directly to the final targets of our business plan. The answer is (drumroll): every division must grow 7 to 12 percent (this is the growth drug I talked about earlier), and, oh, by the way, Mr. Division President, we need your productivity to offset inflation every year, so you need make a 2.5 percent productivity improvement. Don't let it bother you that the building market has been in the tank for three straight years, housing inventory is at an all time high, and the banks are tighter than my high school jeans. The market has stalled, but we are going to perform a little secret to keep our revenue and profits up. This little miracle is called taking market share! Hooray again, everyone, we're saved!

Sound familiar? I know I'm not making this stuff up...

So the division leader gets his engineering, project management, and marketing people together to figure out how he is going to pull a rabbit out of the hat again.

"Where's our list of growth and productivity projects? Great, we've got a long list with lots of ideas. Goody-goody. The marketing people have put their two cents in on the list. Lots of product additions and extensions so we can take that magic market share. Excellent." The project management people breathe a sigh of relief. Here is the magic list that is going enable us to make our plan. *Hallelujah*!

So we have a plan to start with. It's going to be a tough year, toughest one ever, but we have a plan. And, if the Division President can bring home this plan in this tough economy, well, let's just say he can almost taste the COO promotion he has been working on to get himself in the queue for the big guy's job, Mr. CEO himself.

So let's put all this in a spreadsheet and see where we are at against our plan. "Get the sales people on the line to see how much of this great new stuff they are going to sell next year. Let's add it all up and see where what we get. Uh oh, the top line and the bottom line still need work. We are going to need all this stuff and more to have a good plan…"

"Oh, and what sayeth you, Mr. Engineering Manager? Could you please speak up a little," says the Division President, "I'm not sure I can hear your mumbling (grumbling). What was that? We had better apply our engineers against this project plan? Put them in some sort of priority; rank the projects or something? Huh. Why on earth would we do that? We are going to need all this new stuff to make the business successful in this terrible environment. Oh, I see. You don't think you can deliver all of these projects with the number of engineers you have. Huh? Okay. Well, let's schedule another meeting next week, and you come back with a better plan since we really need all of this stuff."

And on and on this goes. You know how this plays out. The divisional leaders have a dozen meetings to work on the plan, and over three or four months time the plan evolves and gets headed in the right direction.

However we still have a gap on productivity.

By the way, by now we have already been in our planning process for next year for six months of this year.... And is anyone having fun yet?

We need to close that productivity gap. So the engineers brainstorm and come up with several other productivity projects that can help the plan meet the bottom line. But by now, there's no way the engineers can complete all of these projects.

The next step is obvious and inevitable. Here it comes. Here we have it, behind door number three, open the curtain, Mr. Division President, and how are we going to fix this gap in our divisional business plan? "Well what on God's green earth are all those engineers in corporate working on, for gosh sake? Let's open up that Pandora's Box to see if those projects are more important than my projects!"

So the Division President gets out a giant pry bar (for leverage—see Engineering 101) and calls in the Director of Corporate Engineering to see what he is planning on working on next year.

Look out, this could be all-out war.

"So what will your fifty engineers in Corporate Engineering will be working on next year, Mr. Engineering Director?" asks the Division President. "I'll show you my big list of important projects. Can you show me yours, Mr. Corporate Engineering Staff Leader?"

"Well, they are working on this new technology for division A, and that advanced product design for your division, and this other new finishing process design that can take cost out of all of our operations and help the environment, and then there is ready-to-serve support for all of the things you don't know you are going to ask me for help with next year, and blah, blah, blah," spews the engineering leader.

"How many of those engineers are working on projects for my division, Mr. Engineering Leader? What was that you said? What is it? Twenty-three? Well, I'll be darned."

And here comes the crescendo…

"Hey, aren't those engineers the same ones that will be allocated to me in my P&L? You remember, ever since we went to our activity-based costing system the divisions are paying for all the corporate resources—the decision we made to charge out all of the shared services, finance, HR, and engineering, to the divisions, so we could have better cost control? Remember? So aren't those resources showing up as *direct* charges to my business now?"

Next comes the dagger to the heart from the Division President. "Hey, those are really my engineers!"

"Bu-bu-but wait a minute" is all that comes blubbering out of the Engineering Director's mouth.

THE LEAN HANGOVER

"Tell you what, buddy, I have a really tough plan this year. Toughest year ever! That stuff your engineers are working on is real, real nice. But I need eighteen of those engineers to close the gap on my productivity and growth projects. Can we work out a plan for something like that? You know my business is the cash cow for the company. The corporation needs me to hit my numbers to make next year's plan. And being the good corporate steward that I am, I'll even let you keep five of my engineers to work on inventing that new stuff. I'll fund you to do that. How's that sound? So you think about how we can do that, and I'll schedule another meeting for next week. Okay? Good meeting. Thanks, buddy."

Here we go again... that Division President does this every year... is all the poor Engineering Director can think to himself.

PART 4

THREE CASE HISTORIES: THE GOOD, THE BAD AND THE UGLY

LESSON 11:

THREE PRODUCT DESIGN STRATEGIES MAKE TOYOTA A GREAT LEAN MANUFACTURER

Toyota uses a disciplined front-end design system to significantly enhance their back-end lean manufacturing capabilities.

I had a unique firsthand experience working with Toyota at their headquarters in Japan, as a supplier in the mid 1990s, and studied what they do in their product design process that makes them the best car manufacturer in the world. Toyota is well known for being the founder of lean manufacturing; however, there are three things that Toyota does during their product design process that make them an even greater lean manufacturer. Even considering Toyota's recent challenges with their recalls for "phantom" vehicle acceleration, I think when you look at these three core processes, you will find the techniques are still best practices in new product design. This is a lesson in how great product design can lead great process design!

First let me set the stage. In the mid nineties, I was an engineering manager for an automotive parts supplier that made wiring harnesses and other electrical components. We won some business with Toyota for several of the vehicles they were building in the United States, and we were building our relationship with them. Toyota was interested in sourcing components in the local markets where they were building vehicles, and we were a competitive US supplier, so we were able to gain some of Toyota's business. Ours was a bit of a complex supply chain; the vehicles were engineered at Toyota's headquarters in Japan and the design information flowed from our design engineers in Japan to our application engineers in the United States and then onto Juarez, Mexico, where the wiring harnesses were being manufactured. Our wiring harnesses were shipped from Mexico to California or Kentucky, where Toyota's autos were being built.

Another factor in this relationship is that wiring is an intensive engineering design effort. Wiring is always one of the last parts to be locked down in the design process because everything else has to be designed first to finalize the wiring harness electrical designs and physical routings. Further, just about every other change that happens to any vehicle subsystem usually affects wiring. Any new options, features, or improvements to the vehicle causes the wiring designs to change as well. It is pretty much a guarantee that with every annual model change, which happen every fall with Toyota, we were going to have anywhere from a few to several dozen engineering changes to our wiring harnesses. So with the high complexity of automobiles today, electrical system design information flow and accuracy were keys to successful new product launches.

One of our strategies to grow a strong relationship with Toyota in this design engineering intensive environment was to propose a co-engineering relationship. We worked with Toyota to develop an agreement that placed several of our engineers directly inside of Toyota's headquarters in Toyota City. These engineers worked side by side with Toyota's electrical engineers to develop the wiring harness designs that would be manufactured in our Mexican plants. This was the same type of co-engineering arrangement that Toyota had with their Keiretsu network of suppliers, who were our major Japanese competitors. However, this had never been done before with an American supplier. At that time we were the first and only American supplier to ever have this kind of access to Toyota's inner workings. We built our Toyota City team into six or seven full-time engineers who were acting just like Toyota engineers with full access to Toyota's engineering and operations. So our engineers got to perceive and experience the "Toyota way" first hand.

This arrangement worked well, enabling us to get our foot in the door with Toyota to show our capabilities and to successfully grow our business. The early to mid 1990s was just a decade or so into Toyota's global market takeover. By 1992 Toyota's cars had already become well known for their high-quality levels and reliability, and their sales were rapidly growing. Toyota's quality levels were the best in the industry, and everyone wanted to know what Toyota was doing to accomplish this. With our engineers inside of

Toyota, doing development work side by side with them, we had a front row seat.

What we learned is that Toyota has a system of design processes that work together to make them an even stronger lean manufacturer. We learned that Toyota uses three key design strategies that fit together to create lean competitive advantage through their lean new product development, as follows:

1. Standardized designs
2. New technologies which are thoroughly developed in research and development
3. Design for manufacturability

You might be thinking that this list looks reasonably intuitive and straightforward; doesn't every business do it this way? Not with the rigor, discipline, and effectiveness that we saw at Toyota. Read on, and I think you will learn why these processes are a key part of Toyota's success.

STANDARDIZED DESIGNS

Toyota has standard, proven designs they use over and over again across all of their vehicle platforms. In fact, they stick to them almost religiously. When we were working with Toyota on the electrical system for a new pickup truck, they had a pre-determined way to do just about everything. Every vehicle subsystem had just one way to be designed, and this was written down on an A3-sized, single sheet of paper that was used over and over again by Toyota's engineers. These A3 standards were the design bible at Toyota. When our engineers needed to work with them to design the cruise control system, Toyota pulled out their standard cruise control A3 schematic. Need to design the anti-lock brake system, just pull out that A3 and use it, and so on.

To understand why this was valuable to Toyota, take, for example, the gauge cluster that contains your speedometer, tachometer, fuel gauge, and warning lights. The industry calls this an *instrument cluster*. When I was on my first assignment as a young engineer working on the J-car platform (small cars) at General Motors, we had something close to fifteen unique instrument clusters across that vehicle platform. In 1992, when we were working with Toyota, their entire vehicle lineup used just two different instrument clusters. Not two just for the pickup we were working on but two for the entire corporation! Toyota made them look different for a Lexus luxury car versus a Toyota pickup, but the important stuff was the same. The guts of how things worked and communicated to the rest of the vehicle—the electrical connections and the circuitry for all of the downstream electrical sub systems—were the same. This was a huge benefit to Toyota's engineering process but an even greater benefit to generating manufacturing efficiencies and to reducing the complexity and performance risks that go along with product design proliferation!

General Motors, on the other hand, had built their business out of leveraging "different" to achieve success. The ladder of success I mentioned previously, Chevy, Pontiac, Oldsmobile, Buick, Cadillac, shared some basic commonality at the chassis level, but GM had gotten into the practice of all too often rationalizing that everything else had to be different in each vehicle for every market segment. In the 1960s Pontiac had a 326-cubic-inch V8 engine; Chevy had a differently designed 327-cubic-inch V8. Go figure. Buick needed a red cruise control indicator; Chevy had a green one, so they had to have different instrument clusters. One division wanted a tachometer, another didn't; add another new instrument cluster.

General Motors had been born out of the conglomeration of six separate brand names under the one GM corporate badge. Each of these divisions had its own separate engineering and marketing groups well into the 1980s, and each of these had rational reasons to do it their own way. And why should GM mess with success; they owned 45 percent of the US market share up until the mid 1980s!

By comparison, Toyota had always been one business under one brand name, with one set of engineers centrally located in their Gijitsu Honkan (engineering headquarters) in Toyota City, Japan. Using proven designs over and over again, across vehicle platforms, was their better business model, and this was a natural development out of Toyota's centralized engineering and central control. What Toyota did was build in differences where they mattered, in styling and appearance, while doggedly protecting similarities in the internal organs of the vehicle to leverage common, proven designs.

You might ask, with all of these common designs what was left to do to engineer a new vehicle? Well, this standardization was only at the subsystem level. There was still all of the work to put all of the subsystems together; package them in the vehicle with all of the dimensions and mountings and connections and then get everything to function seamlessly. That was the work we were doing with Toyota during their vehicle development programs. And while this was a ton of work, the fact that the subsystems were standardized allowed the vehicle packaging to happen more efficiently and in shorter time than the new vehicle development programs going on at General Motors and other car makers.

Toyota used common, proven designs religiously to make product development happen in less time and with more success. They also did this to gain huge benefits across their manufacturing and supply chain operations. The leverage of common designs made the entire upstream system function more efficiently. Standardized designs resulted in less part numbers to source, less to stock, and less to go wrong because they could focus validation and certification on fewer part designs. It also gave their suppliers better leverage and throughput with less diversity and complexity in their operations. Standardized designs increases process capability across Toyota's system to get things done right more often, lowering their overall costs. Standardization is a really great lean product design strategy, and Toyota was using it better than any other auto manufacturer.

NEW TECHNOLOGIES ARE THOROUGHLY DEVELOPED IN RESEARCH AND DEVELOPMENT

The simple reason why Toyota was able to use design standards to create the best vehicles in the industry is that they did their inventing in research and development and not on the critical path of a new vehicle launch. This probably sounds like a no-brainer—do R&D in R&D—but it's still surprising how often businesses put the cart ahead of the horse on this one in an effort to accelerate sales growth.

All of Toyota's new design concepts are proven out in R&D before they are released to the product engineers. This was a firm part of their design culture and business strategy when we were working with them. There are two fundamental reasons why this was important to successful product development. First, it ensures that new technologies are ready for deployment, so the product teams don't take on big risks. Secondly, it creates the organizational discipline to stick to a handful of proven designs, and that in turn creates the design leverage I covered in the last section. Standard designs transcend the individuals who are doing the detailed product designs. Toyota doesn't end up with one group of engineers adopting design A because they think they can prove it out while the rest of the corporation is using design B, for example.

Doing research and development off of critical path is the only healthy way to make sure your new technologies are ready. An eighteen- or twenty-four- or even a thirty-six-month new product launch program is not a good place to try out a brand new, complex technology that you've never used before. That is simply not enough time to do all of the experimenting, testing, and redesign you need to do to prepare a new technology. New inventions almost never work the first time they are conceived and ultimately may even prove to be infeasible. New technologies also need extensive field testing to ensure long-term safety and durability. There is not enough time to do this level of testing while you are on critical path, marching toward a launch date that is only a handful of months away. New technologies typically take anywhere from three to five years to prove out in R&D so that we can place them on the shelf for use in new product programs.

Anyone who has tried crunching down this cycle finds out the hard way that it doesn't work. I've seen this failure play out with young, green, divisional leaders who want to run hard with a new technology so they can commit to the growth the business is demanding—leaders who have not benefitted from the school of hard knocks that comes with putting this cart in front of the R&D horse. There is great pressure in every business's yearly planning cycle to commit to new sales from new ideas, but if the business let's these leaders run with unproven technologies, it is signing up for big risk. Push your new ideas too fast or too hard, before they are ready, and you end up with a laundry list of problems: poor manufacturability, field quality defects, long lead times, and risks in terms of durability, safety, and performance will all result if you haven't done a thorough job developing and validating your new technologies. Be wary of putting invention on critical path; this can quickly compromise your brand equity through poor product performance and delivery challenges.

I must add that the marketing people typically despise this constraint. They want to make the products that their customers need, and they want these products as soon as possible. This is how they close the gaps on their annual sales plans and get an excellent rating and a big raise on their annual performance reviews! Design rules and waiting for R&D puts these ideas on hold. However, if you let the marketing people push your technology programs ahead of your R&D you will get poor results. Leave real development to R&D.

This is pretty basic thinking, so you might be thinking, *Why does this happen?* It happens in businesses that don't have the foresight or discipline to commit to R&D as a long-term investment in their business. It takes top-level discipline to say no to launching a new idea that hasn't been properly incubated and to instead develop the R&D processes necessary to feed the pipeline, especially in businesses that drive for strong productivity, continuous improvement, and growth every year. Businesses have to have a long-range product planning vision that is separate from the mainstream to identify and deliver on good R&D programs.

At this point you may also be wondering what R&D off critical path has to do with Toyota's lean manufacturing system. Simply put, Toyota could not have had their A3 designs standards ready for deployment without this R&D discipline. If we allow new technology development to happen on the critical path of new product development, we get whatever we get. When the development program yields unexpected results, new product programs will push forward with riskier kinds of solutions. We end up with compromised designs that arise program by program instead of proven, common designs that we can count on. Toyota has been able to create both design and manufacturing leverage through these disciplines. They develop and stick to proven product standards, and the reason they have these product standards is that they do their homework in R&D to prepare new subsystem designs.

DESIGN FOR MANUFACTURABILITY (DFM)

There are plenty of available resources and literature on the details of design for manufacturability, but, in brief, DFM is the philosophy of value engineering product designs to make them easier to manufacture by taking parts count down and reducing labor content and manufacturing operations through simpler product designs. As an example, brute force engineering uses a screw and a nut to hold everything together while elegant design uses snap-together pieces to hold parts together. Another simple DFM example is integrating a metal bracket into a plastic-molded part to combine the two parts into one. Significant improvements can be made when focus a group of engineers on simplifying product designs.

Early in my GM career, when Toyota was making huge sales gains in the United States, we performed vehicle teardowns to learn what Toyota was doing to achieve their low costs and high quality. We laid out all of the parts from a given Toyota vehicle on one set of tables alongside tables containing the same parts from comparable GM vehicles. The results of these competitive benchmarking exercises were astonishing. I can remember the Toyota vehicles requiring less than half the number of tables to display their parts, with Toyota having as few as one-third to one-half the parts of the comparable GM vehicle.

This was over twenty years ago, but I can remember thinking to myself that it looked like we needed a complete do-over at GM. I was standing there, trying to think logically how GM was going to be able to survive long enough with our high cost, overly complicated designs, and to be able to unwind all of that complexity to catch up with Toyota. It looked like Toyota had started their company under a different set of rules, and better rules, at that. And, in fact, they had done exactly that. There were years of capital investment and design momentum baked into our GM designs, and it wasn't apparent how we could turn that around. We, the GM design teams, were going to have to have enough time and discipline to improve our new vehicle designs under this new set of competitive rules. Looking back now, it took a couple of generations of designs for GM to significantly catch up, but for the most part they appear to have barely survived this challenge.

Simpler designs are another powerful advantage that Toyota uses to make their business leaner, but how did they actually become better at this, and what does this have to do with their lean design system?

Here is what we experienced. When engineers design parts in general, they work to get the system level functionality done first and then worry about the packaging later. This makes intuitive sense; when engineers are tied up solving functional problems during the design process, they leave packaging for much later, after the base design is settled in. And the easiest way to package a part is to add on a bracket and some screws or bolts, and that is exactly what happens when designs aren't settled in early enough and crunch time comes for finalizing designs.

What I saw at Toyota is that they built a system that got the functional designs right early on via their A3 design standards. That way they could dedicate more of their vehicle development time to nailing down all of the dimensional relationships in the vehicle and doing more work to refine the manufacturability of the final designs. Toyota was using several levels of prototyping to exercise new designs and refine them to get ready for launch. Design for manufacturability was part of the agenda and not an afterthought. Remarkably, and a bit counter intuitively, they were even doing all of this better work in less time than any other vehicle manufacturer all because their *total design system* was planned to be more efficient.

Toyota is known for of its development of lean primarily as a manufacturing tool. Through our co-engineering relationship I was able to experience how Toyota's lean design capabilities also add to their overall lean business strategy. When you step back and look at these three design capabilities in combination—design standards, disciplined R&D, and design for manufacturability—it was clear Toyota was also using their front-end design system to significantly enhance their backend lean manufacturing capabilities. Toyota was creating a system of processes in both design and manufacturing that worked together to make them the greatest lean manufacturer ever!

These are some great lean lessons as we think about the reach that our product development can have on a lean business.

LESSON 12:

MISDIRECTED LEAN NEARLY KILLED THIS FURNITURE BUSINESS

> Pulling the levers on small batch sizes too early in a lean transformation can crash shop floor flow and send the business into chaos.

Tom Carver had been supporting the Wright Furniture business from his corporate role for several years when he was offered the vice president of operations role at Wright. Tom knew Wright Furniture was struggling. It had made some attempts at lean manufacturing, which were not yielding positive results, and the business had recently taken on a difficult change in its information system in response to the Y2K threat, an *enterprise resource planning* (ERP) system implementation. He also knew with trepidation that he was going to be the fourth VP of operations at Wright within the last ten years. Tom sensed this opportunity was going to test his capabilities, but he felt he had the capability to help improve the business. Tom was a man of resolve, and he felt he was up to a new challenge.

Wright is a one-hundred-year-old high-end furniture manufacturer, marketing to the top wage earners in the United States. Wright makes high quality wood case goods and upholstered furniture for the most discerning tastes. Wright Furniture was well known for creating some of the finest original and period reproduction furniture available anywhere in the world, and they were proud of the strong brand position they had earned with high-income earners and the best interior designers around the nation.

The business had several factories in the United States and a factory in the Philippines. The furniture industry had been moving production to Asia for some time, due to low labor costs and the availability of local woods like the mahogany grown there. Wright was no different; they manufactured some of their simpler goods at a plant in the Philippines and also sourced roughly 40 percent of their products from other suppliers in China, the Philippines, and Indonesia. The result of this structure was that Wright's domestic factories were challenged more and more over time with making only the complex, higher-end pieces. This provided a tough environment for the US plants where costs continued to rise as a result of being focused on manufacturing only the complex end of the product line.

Wright's business also had a rigorous new product introduction cycle. Most furniture businesses display their new lines twice per year at the spring and fall furniture markets in Highpoint, North Carolina. Buyers come to these events from all over the world to see the newest in furniture designs and to place their factory orders for the upcoming selling season. Wright furniture had always been the best in the industry, and Wright's customers expected to see the freshest, newest, and most stunning furniture at each market from them. Each furniture market also brought a fresh new offering of fabrics in the latest colors and prints for Wright's upholstered furniture, and the business might also add a new wood finish or two to the lineup. So every six months, Wright Furniture had to develop roughly thirty to fifty pieces of new wood furniture and upholstered goods to keep the business performing.

Wright was at the top of this food chain, offering more new innovative fashions per year than any competitor. As a result, Wright's manufacturing operations were constantly challenged with complicated new product introductions, replacing roughly one third to one half of their product line every year. Further, in the furniture industry new product development is a lot like throwing spaghetti at the wall; when the spaghetti sticks, it's ready! The majority of Wright's furniture designs never sold very well; average total lifetime sales across the product line was on the order of one to two hundred pieces. And, many of the designs never sold more than a dozen or two pieces. On the other hand, a very few of the most desirable designs did sell well, taking off immediately after these market introductions. The take away is that volumes on new product introductions were very unpredictable, challenging operations to guesstimate the demand for each new design.

Tom Carver hit the ground running. During his first days on the job, Tom spent some time with the outgoing VP of operations, Billy Van Zee, in an effort to create a smooth transition. Tom and Billy walked through each of the US factories, meeting the team and going over the initiatives Billy had been driving and where Billy saw the major challenges. Billy was a shrewd, shop-floor educated leader, yet he knew his efforts at Wright had been less than successful. Tom and Billy's time together was cordial, trading war stories on the difficult challenges in the furniture business and meeting with the team to show that a unified transition was in place.

However, it didn't take long for Tom Carver to start to get a sense for what was going wrong at Wright. During the walk, Billy Van Zee talked about the efforts he had made developing his lean manufacturing initiatives and that even a couple of the case goods plants had installed "lean learning lines." These were one-piece flow cellular lines where small batches of components could be assembled to make one specific model of furniture. This was interesting to Tom Carver and he was encouraged that the factories were learning about lean. However, Tom was also skeptical about how these lean learning lines were ever going to apply in a general way across the business with all of the product turnover and unpredictable sales rates that were a regular part of Wright business.

In addition Billy relayed that he had centralized scheduling for all of the case goods manufacturing and sourcing under Joe Wilson, Billy's new supply chain manager. Previously this scheduling had been done locally within each plant. Rick Plow was the actual production scheduler reporting to Joe. Other than the lean learning lines, this business was still one hundred percent build to inventory with a batch-and-queue manufacturing system. Rick's job was to analyze demand rates versus inventory on hand to decide when he needed to schedule each new batch of furniture to be manufactured.

The other interesting thing Tom Carver learned from Billy was that Billy had been cutting batch sizes in an effort to lean out the business. Billy managed this process through this new central scheduling team. Historically, furniture factories had built big batches to keep their costs down, amortizing machine setup and assembly setup costs over these large batches. Dining chairs would typically be built in batches of two hundred fifty to five hundred units. Dining tables and dressers might be built in batches of seventy-five to two hundred fifty, depending on the demand rate. Small batch sizes, less than roughly fifty pieces, were rarely ever used because of the high setup costs to produce furniture parts.

Billy's direction had been to drive batch sizes down in two phases, first taking them down to fifty-to-one hundred range and then down even further to the twenty-to-sixty range. Billy explained that this was his way to push the plant managers to figure out lean. Tom's antennae were up, hearing this information, and light bulbs were starting to go off. Things were starting to smell suspect with Billy's lean directions.

The next thing Tom Carver decided to do was to walk the shop floor to "be a furniture part for a day." Furniture is generally made from two groups of wood parts: solid wood components, like legs or rails, and engineered wood panels, like veneered tops and case sides. Tom wanted to understand what was really going on in the plants by walking the routings of a couple of typical furniture parts. Frank Beeley, one Wright's best design engineers, volunteered to walk Tom through the routings. Tom asked Frank to help him walk a solid wood sideboard leg and curved plywood drawer front to see the range of Wright's manufacturing processes.

What Tom learned through his walk was concerning. Wright's new ERP computer system had been implementated about eighteen months ago. Prior to this ERP conversion, Wright Furniture had run their factories through a system of manual part routings.

In the old manual system, when a piece of furniture was scheduled to be produced, the routings for all the required parts were printed off on colored pieces of paper and then dropped off at the starting machine for each piece. For example, when a dining table was scheduled for production, each of the component routings for that dining table would be printed off on an orange piece of paper. It was then the responsibility of the operators to move the parts through the machines following the order within the machine queues. Each operator's job was to move as many parts as possible through each machine on a daily basis, and the colored pieces of paper went with each part.

Wright furniture's production control system was supposed to be driven off of the final assembly dates. When furniture components were completely machined and ready for final assembly, they were individually queued up in a defined location on the shop floor. The materials management staff would monitor the upcoming assembly schedule and work with production to manage component flow through the system to have all of the parts ready for final assembly on time. However, and as was normally the case, most assembly runs required manual intervention to get the all of the required parts to the assembly area on time. Wright's material expediters were responsible to push the required components through the machining operations, and the colored sheets of paper were used as visual identifiers to aid in the process. The expediters would walk the shop floor, looking for the particular color routing sheets to find components that needed to be pushed through the machines to get them to assembly.

Wright's batch-and-queue shop floor was clogged with parts and assemblies everywhere. Under Wright's batch-and-queue manufacturing system, parts were rarely ready for assembly on time. Normal part routings varied from a couple of days for a simple component to a couple of months for some of the more complicated parts. Even when things went according to plan, fully machined parts would show up in the assembly queue at random and uncoordinated times because of these differences in part routing times.

Wright's new ERP system was supposed to improve this. It would control part releases to the shop floor according to computerized routings and timings so that a part that takes five days to manufacture would no longer be started at the same time as one that takes twenty days to process. The engineers would feed the ERP system theoretical part processing times, and the system would coordinate overall timings and flow. The theory was that parts would only be on the shop floor when they actually needed to be, thereby unclogging the machine queues. In this new system parts should flow through the factories more consistently and closer to plan.

Unfortunately, what Tom found out in his parts walks was that the Wright team did something really creative with their ERP implementation. The Wright engineers decided that they wanted to be sure that furniture parts had enough time to get through the shop floor, so they added a day of wait (queue) time in the ERP system for every machine operation that a part would go through. Their thinking was that parts never used to get to assembly on time, so this was a planned improvement to make sure parts would have enough time on the shop floor to ensure reliable production flow. This way all parts were *guaranteed* to get to assembly on time!

This may have been reasonable thinking on the surface, but the actual impact of adding all this planned wait time to parts was not good. In fact, it crippled the business. The drawer front that Tom was walking through the shop floor was complex and had many machine operations. It had a very fancy veneer face that was laid up from several intricate laser cut wood pieces made out of several different wood species. This veneer face took a couple of weeks for Wright's internal veneer plant to make. Then the face had to go to an outside supplier for two to four weeks so it could be glued up into the curved shape of the drawer front. Then it came back into Wright's operations for a long list of machining operations. The part took too many days to process before the ERP system implementation. However, after the ERP work, the team had engineered this part routing so that it now took 126 days to move through the shop floor! That is 126 working days, or roughly five months! *Incredible,* Tom thought to himself, *this is a very nice drawer front, but… we are dead… there is no way we can take five months to manufacture our components and stay competitive!*

And, as one last supporting piece of information, as Tom was doing his "be a part for a day walks," he observed a machine operator doing a setup on a tenoner. Tom asked the associate how the setup was going and how long it was going to take. The operator said that it usually took him about an hour to do a setup, but this particular part was fussy and was taking a little longer. Tom then asked the operator how long it would take to actually run the parts once the machine was ready. The man chuckled a little as he relayed to Tom that it would probably take only a couple of minutes to run this batch of parts through the machine. Tom noticed a couple of parts laying aside that the operator had used to test the setup. These test parts were ready for the scrap pile. Tom knew from his days in corporate that long setup times had always been a major challenge in the furniture industry, and this is fundamentally why furniture businesses manufactured in large batches. This last stop confirmed to Tom that Billy Van Zee's smaller-batch-size strategic direction was having a huge negative impact on cost and throughput in the business. The setup time on the tenoner was too long to effectively process small batch sizes. And throughput was being further exacerbated by the impact of losing a few pieces to setup trials out of an already small batch.

It was pretty clear to Tom now why the business was in trouble. The challenges at Wright furniture were starting to unfold, and his walks provided the "aha" moment Tom Carver was hoping he might find. It was no longer a mystery why Wright Furniture's throughput had crashed and costs were skyrocketing.

First Billy Van Zee had reduced batch sizes to levels that were too costly for the business. The batch sizes that came out of Billy Van Zee's lean initiative were causing the long setup times to be "absorbed" by only the few pieces that were being run in these small batch sizes. The machines were spending more than 50 percent of their available production time in setup because of the short processing time for such small batches, and killing machine throughput. This was not a good way to run a factory.

Secondly, the ERP implementation had added insult to injury with over inflated parts processing times. The lead times designed into the new ERP system were way overstated; extending Wright's historically long lead times even more and, worse yet, further clogging the shop floor with inventory.

Thirdly, parts lost during setup trials became even more impactful on throughput and cost with Billy's small batch initiative. Trial parts were always part of the business, but they had historically been a low percentage of necessary scrap within a large batch. Losing three table tops out of a one-hundred-piece batch during a machine setup generates a 3 percent loss, losing the same three table tops out of a thirty-piece batch, now created a 10 percent loss. And every one of these furniture parts goes through multiple machines setups, so in Wright's current situation as much as 50 percent of the batch was being lost to setup trials.

The manifestations of these miss-directed lean directions on the business were severe; Wright's costs were skyrocketing, throughput had crashed and all of the stress was having an impact on quality. All of their wood parts sitting around the shop floor for so long would allow them to crack and warp and turn bad, so quality was suffering too. Wright's ERP and lean directions were basically crashing throughput, driving costs up, and outright killing the business.

Not surprisingly, field quality problems had also risen drastically over the past twelve months at Wright. Returns were up across the business due to the pressure these issues were putting on production to keep up with demand. So, while Tom was absorbing all of the challenges in the business, he decided to take one more internal temperature check. Tom decided to perform a random quality audit of finished goods pieces to see what was going on with Wright's quality and brand commitment. What Tom found was that quality out of his newest plant in Lewisville, North Carolina, was terrible. Quality was bad on seven out of ten items he inspected from finished goods inventory in the Lewisville factory.

Lewisville was the newest plant in Wright's operations, and the plant had had an accelerated lean model, one where the plant manager had attempted to create flow with his machine placements and cut batch sizes the deepest. As a result, Lewisville was struggling to get good products through the plant and apparently shipping whatever it took to make their production numbers.

Lastly, and piling on, Tom also discovered that roughly two thousand pieces of furniture were stuck in Lewisville's finishing room where the stains and lacquers are applied. This was a plant whose output at that time was only about thirty to forty pieces per day. Tom discovered Lewisville had two to three months of production output stuck in the finish room because of quality problems that could only be detected once the finish was applied. The bad pieces had been sent back to the finish room to be stripped down, repaired, and then refinished. The Lewisville plant manager was really struggling with his operations.

The depth and severity of these challenges was a stark surprise to Tom as he tried to get his head around how he was going to get this train back on the tracks. It didn't take more than these first few days for Tom to understand how bad his situation really was. It appeared to Tom that the business had nearly lost the recipe for successfully making make fine furniture. Wright furniture had a perfect storm on its hands, and the business was falling into a death spiral.

So let's stop for a moment here and check Wright Furniture's lean conversion scorecard at this point. Here are the results of Billy Van Zee's lean improvement program: costs up, quality suffering, throughput crashing, and customers running for the doors! Not exactly what a business would normally expect to gain from lean manufacturing. Chalk a win up for misunderstanding lean and also for misunderstanding the application of an ERP system!

This was not a productive time for anyone at Wright, but fortunately there were logical solutions. Wright had two main failures that had combined to cause just about everything to spiral downward in the business. These two root causes were (1) the business took batch sizes down before its processes were ready in a miss-directed lean conversion, and (2) the business had programmed overstated routing lead times into its ERP implementation.

Tom had to make some changes fast before the business fell into complete financial ruin. He had to figure out how to uncouple this mess and set some short-term tactical directions, and he needed a good long-term plan coupled with hard work and fortitude to turn the business around. Tom realized he had to get Wright back to stability and stop the bleeding before he took on any longer term lean initiatives.

The first thing Tom decided to do was to assemble a team that could better lead the business. It was clear several members of the leadership team had checked out due to the poor working conditions in the business. Tom had to make the tough decision to let two plant mangers go, including the Lewisville plant manager. Tom also needed stronger talent to lead production planning, someone who understood lean in an ERP environment. He coached Joe Wilson and Rick Plow out of the business and put in a young engineer who understood how to production plan in an ERP environment. Tom also needed an engineering leader and a couple of process engineers to fix the ERP part routings short term and to develop stronger lean processing solutions for the long term. He hired these. Tom's most experience leader, Wright's international sourcing manager, left because he understood how bad things were. Tom had to find a replacement for him ASAP. All of this occurred over just a two to three month time frame. It was a challenging time for Tom and the entire team at Wright.

Tom Carver started his improvement process in the production scheduling area. He decided to meet regularly with his production planners to help quarterback the scheduling process. Tom needed to take his batch sizes up quickly and take the ERP lead times down, so these new batches would start to flow more efficiently. Even with those changes, Wright wasn't going to see significant output increases for many months. The overstated lead times in the new ERP system had created months and months of work in process on the shop floor. It was going to take time to flow these small batches through the system to finally get the new big batches into production.

Wright had also dug a hole with finished goods inventory levels because output had been so poor for so long. The decrease in throughput had caused the company fall behind incoming orders. Inventory was way behind demand for most popular skus, and Wright was losing ground daily.

Tom's strategy to increase batch sizes was going to use Wright's production capacity over fewer skus in the near term, meaning they were going to only be able to build certain products over the next few months. These changes were going to result in a better business long term but pretty much guaranteed poor service levels for the next several months.

Tom also found out through these production scheduling meetings that, under the pressure of Wright's production output crashing, Rick Plow had been adjusting the timing and priorities in the build schedules on the fly. If Rick Plow thought that Wright needed one product over another, he adjusted the assembly schedules to pull ahead items with larger stock outs. A small degree of this could be manageable, but Wright had such a deep finished goods inventory shortfall that Rick's schedule changes had gotten out of control. Early on Tom Carver compared one month's schedule to the next, and the manual alterations were stark; over 40 percent of the planned assembly dates had changed. Wright was chasing its tail, trying to react to the widening gap in its ability to meet delivery dates with frequent final assembly schedule changes.

What Tom and his production scheduling team decided to do for their new plan was to scrap out some of the smaller batches that were early in their manufacturing lives. This was a short-term financial hit, but it was a way to free up needed capacity for bigger batches. These small batches were going to cost more money to make than they were worth, anyhow. Tom started dropping in bigger batches as soon as he could and concurrently had the engineers take the ERP lead times down. Tom monitored the production plan every Monday to work through these tough times.

Tom Carver also decided to take the focus off of the one-piece flow lean cells that had been put in place by Billy

Van Zee. These were interesting theoretical exercises and might have been good training grounds, but they were not the right focus for the business at this time.

The business had a ton of substandard quality inventory in the warehouses and in the factories. The bad pieces coming out of Lewisville had become so prolific a few employees in cahoots with a security guard had setup an outside business refurbishing and selling the pieces that the factory was sending to the scrap pile. The warehouse team isolated the bad inventory from the warehouses and from the shop floor and moved them to a warehouse to sell to employees and their families at heavy discounts.

Wright Furniture was taking financial hits left and right. Fortunately the mother ship had deep pockets, or at this point Wright would have probably been bankrupt had it been a standalone business.

It always takes time for changes to flow through the system and prove out, especially in an ultra-low inventory turns business like Wright furniture. The leaner you are, the quicker a change flows through your system. Unfortunately the opposite is also true. Tom felt that he was doing the right things, but he knew it was going to take several months to show in the business's financial results.

Tom met with the owners of the business after two months on the job to talk them through the situation. Remember, Tom was the fourth leader the owners had seen in this role over the past ten years, and every one of them had failed to fix the business. Tom tried to explain to them that he could get Wright Furniture back to stable production within six to nine months but that they were going to have to take some write offs to purge the sins of the past. Tom tried to use the analogy that turning this business was more like turning a large supertanker where you have to start miles ahead, compared to turning a ski boat that can turn on a dime.

Business owners want results, and they want them now. All Tom's CEO could relate to was that Wright had tons of capacity and had an over abundance of plants and people that the corporation had invested in. Why couldn't Tom use these immediately to make what he needed to make? The CEO made a convincing argument, and in some ways, he was right; unfortunately it wasn't possible for Tom's changes to happen this quickly once you got down into the details of how the changes needed to be done. It was a rough meeting with some fist pounding and such, but Tom survived to fight another day.

Tom was scheduled to be back to corporate headquarters in four months for another update. By the time this second meeting came around, things had started to improve at Wright Furniture pretty much to the plans and the commitments Tom had made. Tom's staff had even begun to work like a team. It was the first time in a couple of years that Wright's leadership team acted like they were all in the same boat rowing together.

This second meeting was an interesting dynamic that went something like this. Tom was still a bit anxious, as one gets for these big reviews, with his job literally on the line. Wright's financials were not stellar, but they had improved drastically, and the business was no longer bleeding red. The business owners had extensive monthly financial reports, so they knew exactly where things stood well before Tom walked into the meeting. The meeting started slowly and after only ten or fifteen minutes of small talk on these operational performance improvements, the attention suddenly shifted. One of the corporate leaders noted that with this strong improvement in manufacturing and now that manufacturing was obviously "fixed," why wasn't sales stepping up to the plate, and why weren't they out selling harder? Now that the Wright brand was back and manufacturing was fixed what was the sales team waiting for?

There was a silence on the side of the Wright business team. This was totally unexpected, and Tom was stunned. Sales had been totally stranded by manufacturing for the past eighteen-plus months. Tom knew this, and he, as much as anyone, felt for the sales team. Sales had been keeping the business afloat by schmoozing key customers and convincing them Wright had identified the issues and made changes and had things under control with quality and delivery and that better days were just around the corner.

It was one of the strangest meetings Tom had ever experienced. Tom hadn't slept well for nights leading up to this meeting because his job was on the line. Instead of Tom taking the pounding, the meeting resulted in Wright's VP of Sales back on his heels trying to respond to why his sales were so poor. The sales leader didn't deserve it, but for Tom it was a win to have some of the heat off of manufacturing for now at least.

This Wright Furniture example is really about how not to misfire with your lean directions. These errors in management direction seem sophomoric when we put them down on paper, but they were not so simple while they were playing out for Tom Carver and his business. It's easier than one might think to turn the knobs too early in the small batch direction with lean and upset the balance of flow in manufacturing. Wright sure had done it with its batch sizes and ERP implementation. Smart, capable, experienced people in the business had made these decisions that caused the business to crash. Internal and external customers thought Wright Furniture had lost its recipe for making furniture, and in a sense they were correct.

Wright was only getting back to stability in manufacturing with the corrections that Tom had made. They had stopped bleeding red on the P&L, but, even after these improvements, Wright was still only a marginal business plagued by long lead times, high costs, and poor profitability. They had scratched their way back to being a traditional, poor performing batch-and-queue manufacturer!

So, longer term, was there a valid lean program for Wright Furniture?

Yes. Wright absolutely needed lean to turn their business into a strong performer, and the team hadn't given up on its lean journey. Wright had huge opportunities for lean everywhere. In fact, what they needed to do had become even clearer with the lessons of the past. It was just a matter of figuring out the right lean strategies for the business.

Once Wright Furniture was able to get its head above water, Tom started developing a real lean plan for the business. He started with his largest case goods factory, and, using a one-week kaizen event format, Tom was able to work with his key leaders to lay out a vision of what they wanted the plants to run like as their first attempt at viable lean. It consisted of over a dozen areas where Wright could put machines together in cells to create common flows. These cells would be designed to process parts in continuous flow with little or no changeover to take lead time and cost out.

Tom knew furniture still had lots of parts that don't flow commonly, so he also envisioned having a "spaghetti bowl" of machines where these unique complicated flows could be isolated and managed better. The team was thinking lean, with common flows where possible, and isolating complexity where flow was not possible.

The team's vision was right on. These were the changes Wright Furniture needed to make to be able to ultimately take batch sizes down and turn this into a shorter lead-time business. This was a long-range plan that would take many years to complete, and it was likely to change and evolve as time went by, but it was headed in the right direction. The changes would enable Wright to take significant cost and lead time out of the business. It was going to take a lot of hard work and organizational change to get it done and Wright needed to build substantial product and process engineering capabilities to have any chance of making lean successful.

They started off by building a dedicated cell for assembling dining tables. It worked well. Wright's lean program was out of the gates and on its way. It felt good for Tom to know what his team had accomplished to bring the business back to life and to finally undertake a lean journey that was based upon good strategic direction. The catastrophe caused by Wright Furniture's poor past lean directions and ERP implementation was history now. Tom knew he had a lot of work left to do at Wright Furniture, but he was confident that he would be successful driving the lean program now.

LESSON 13:

CABINET BUSINESS: EVOLVE YOUR LEAN OR DIE

Early lean manufacturing models often need to change and adapt as the business grows.

Let's take a look at a kitchen cabinet business where lean seemed to get off to a good start, but static, early lean decisions ultimately led to the failure of the business. We can use this example to learn about some of the risks that come from not evolving a lean system with a growing business.

High-end kitchen cabinet companies can be very complex build to order businesses, mostly because of the breadth and width of their product offerings. A large cabinet company may offer anywhere from twenty to forty different door and drawer styles, from traditional stile and rail doors to modern mitered corner doors to flat panel arts and crafts style doors to arched panel doors. Cabinet businesses also offer a host of cabinet models with wall, base, vanity, pantry, bookshelf, entertainment, and others normally in the lineup. And cabinet businesses must offer a host of sizes. Standard widths are offered in inch or half-inch increments, and when those don't fit just right, most cabinet companies will build any custom size needed.

Continuing with this complexity, cabinet businesses will also build their cabinets in several different wood species and typically in twenty or more different paint and stain finishes. Then there's face-frame and Euro-style box designs, a range of drawer slides, glazed and faux finishes, and finished plywood cabinet boxes versus particle board boxes. Cabinet companies create sales through a host of differentiation and customization to meet personal tastes and architectural needs. A cabinet business becomes, at the core, the business of complexity management.

As a result of all of these product offerings, most custom and semi-custom cabinet companies build their products to order. It's nearly impossible to build cabinets to inventory because there are too many sizes, styles, finishes, and other variations to ever build anything reliably to a forecast. In essence, every order that comes in is a custom order, never to be reproduced again.

The strongest cabinet companies have become very proficient at managing all of this complexity such that they can build to order in very short lead times. Many cabinet companies have become world-class lean manufacturers, turning around their orders in five to ten days' lead time. So in the cabinet industry, every order is really the equivalent of a fully custom order that must be built and shipped in just a few days. Today's custom and semi-custom cabinet businesses must be run as build-to-order operations with very short manufacturing lead times to be competitive.

Cabinet factories offer one of the best business models for meaningful lean manufacturing—continuous flow manufacturing fits the shop floor like a glove. Cabinet businesses have a consistent pattern of process steps, end to end through the factory and even with the great product variation that was covered above. And the process flow remains relatively simple, with a consistent, linear, stepwise A/B/C/D style of process flow. This allows for relatively straightforward alignment of machines into cellular flow processes. Many cabinet businesses have been able to effectively use lean manufacturing to generate reliable, high quality, short lead times.

Our example company, Rowe Cabinets, was originally started by Bill Rowe, who built the business with his personal assets and sweat equity. Bill understood the cabinet industry well, and he understood the need to build to order to be successful with his business. Bill hired Frank Jackson to head up his operations, and Frank developed into the key manufacturing thought leader in the business. Together they developed a very successful semi-custom cabinet business that could build cabinets to order and deliver high-quality products in world-class lead times.

Following is the basic formula that Bill Rowe and Frank Jackson developed as their manufacturing strategy:

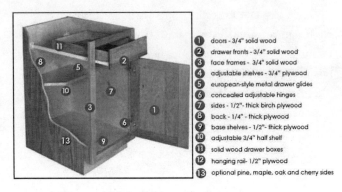

Basic Cabinet Design

1. Cabinet boxes are cut and assembled to order within the Rowe factory.

2. Doors and drawer fronts are sourced from a local supplier and stored in a grocery store style inventory system. Only standard drawer and door front sizes are stocked to make inventory levels manageable. Custom sizes are built to order at this same supplier with expedited timing.

3. Dovetailed drawer boxes are built from a grocery store of drawer sides. An outside vendor supplies the drawer box components. Custom-size drawer boxes are also built to order at the local supplier through expedited timing.

4. The business was designed around a strong machine build capability, which manufactures and maintains Rowe's custom machines, jigs, and fixtures that are integral to Rowe's continuous manufacturing flow processes and short lead times.

5. Orders are built in a seven day lead time as follows:

	Doors/Drawer Fronts	Cabinet Boxes	Drawer Boxes
Day 1	Order custom sizes from supplier	Load cut panel sizes	Order custom-sized components from supplier
Day 2–3	Wait	Cut panels	Wait
Day 4–6	Receive custom sizes; pull standard inventory; sand and machine for hinges and hardware	Assemble boxes	Receive custom-sized parts; pull standard inventory; assemble drawer boxes
Day 6	Apply stain and clear topcoat	Apply clear topcoat	Apply clear topcoat
Day 7	Assemble, pack, and stage for shipping		

Rowe Cabinet's manufacturing strategy was fairly straightforward and it created a very successful cabinet business. The manufacturing strategies Bill Rowe used were also very logical, ones that any one of us might choose if we were going to recreate this business today.

Let's drill a little deeper on a couple of these decisions to better understand Rowe's thinking. This will give us the background to understand how these decisions caused the business to struggle later on. Bill's decision to outsource cabinet doors and drawer fronts made good financial sense for his start-up business. There is a ton of door capacity in the cabinet industry; just look through an issue of *Fine Woodworking* magazine, and you will find several potential suppliers. For a small start-up business like Rowe Cabinets, this was a great way to avoid investing in all of the machines needed to make cabinet doors.

The decision on drawer sides is also a good direction for a start-up business like Rowe Cabinets. There are many capable drawer side and drawer box manufacturers that focus on just this niche. Cost-effective businesses exist that just make drawer sides out of cheaper, reclaimed lumber scraps from higher-grade wood industries. The wood for drawer parts tends to be a second-grade wood since it is not part of the cabinet exterior. Drawer sides can be made up of thinner, short pieces of wood with poorer grain and coloring—wood that is considered scrap for many furniture grade applications. Many drawer manufacturers get their wood for very low cost because they use cutoffs from other wood businesses. So this was another good decision for Bill Rowe's small start-up business.

Bill's decision to outsource cabinet door and drawer parts provided focus and reduced human resource and capital investment needs for his young business. Further, these strategies were lower-cost alternatives compared to making the doors and drawer sides in house. Rowe's suppliers had volume leverage and focus so they could make these components at lower costs than Rowe could themselves.

And, Bill Rowe also had excellent insight to invest in an internal machine design and build capability, especially since he was in a wood-working business. The need to have an internal machine build capability to create a true lean business is one we discussed at length in Lessons 2 through 4. Because wood can be easily machined and joined into unique shapes, wood industries tend to have a higher variety of offerings. Custom machines that shape, dimension, and profile these parts go hand in hand with the business, as do jigs and fixtures for machining, assembling, and joining wood parts. The ability to create fast, efficient flow on the manufacturing shop floor can be done better through custom process and machine designs.

So far, Bill Rowe's manufacturing decisions seem to be yielding all positives for his business. And these strategies did prove out to be good early directions for the business. The business was capable of managing a reasonably broad product line, consistently manufacturing high-quality cabinets in two week's lead time or less. Thus, the business was successful and grew rapidly. In less than fifteen years, the business had grown to annual revenues of over one hundred million dollars.

The business's rapid growth and successful manufacturing formula also proved attractive to potential investors. The construction industry had been on a lengthy upward trend, which was bringing high market values for strong buildings products companies. Bill had created a profitable business model that could make him financially secure if he sold his business. So, Bill decided to put his cabinet business on the market and, after a lengthy process, the business was sold to a large, privately owned building products company. The new owners were particularly attracted to the business because of Rowe's lean manufacturing capability and how that had brought rapid growth to the business. The businesses that made up this new corporate owner were primarily long lead time, batch-and-queue businesses. The corporation had dabbled in lean manufacturing with modest success but had not made a commitment to lean. The new owners had paid a premium for Rowe Cabinets because they felt they had purchased a higher performance lean business, one that they planned to leverage by transferring this lean know-how to other parts of the business.

Rowe cabinets ran along fine for some times after the acquisition. It was pretty much business as usual. However, as a few years passed, things started to change in the business. The new corporate ownership pushed to grow the business to recoup its premium investment, and as a response Rowe expanded its product offerings to sell more. Mitered door styles were added to the traditional coped rail and stile doors. More wood species and finishes were added to continue to drive growth. Contemporary cabinet shapes with curved fronts were added to lineup. Distribution was added to more distant parts of the country where Rowe had never successfully done business before due to overly high transportation costs.

The changes were imperceptible at first, but Rowe's longstanding high quality and short lead times slowly started to degrade. The business was starting to feel some pain from adding all of these new products and distribution and the resulting growth. Complexity multiplied quickly in the factory with all the product additions. Higher sales meant more of everything going through the factory, and operations were starting to be pushed to their limits. Growth is what every business works hard to achieve, but as a business grows, it also needs to reevaluate and evolve its manufacturing strategies. The successful strategies of a small start-up company may not prove to be effective as the company becomes larger.

This was exactly what that was happening at Rowe Cabinets. As all of this complexity and volume piled on, Rowe's early lean manufacturing strategies started to create new challenges. Rowe was still sourcing cabinet doors, drawer fronts, and drawer sides at this point and still stocking them in an inventory grocery store. They had been doing this successfully since day one. However, with the size of the business now, and with the more diverse product lineup, inventory grew to unmanageable levels for these stocked components.

At its peak, the factory was making fifteen hundred to two thousand cabinets per day. The inventory required to reasonably service this level of business across all sizes, styles, and species of doors was nearly impossible to manage. The business had door and drawer front inventory stacked in every nook and cranny of the factory. The shop floor became a mess, with tens of thousands of these components stacked in racks and cubbies that ran from floor to ceiling, nearly everywhere the eye could see.

Even with these stockpiles of inventory, stock outs and expedites became more prevalent. There was no way Rowe Cabinets could stock enough of the right inventory at the right times with their diverse product offering. Rowe could only afford a limited stock of any given size and style of door and obviously had limited space. Frequent stock outs were inevitable with this operating model, and the business was putting more pressure on their outside suppliers to expedite more and more components to order, in short lead times and ever increasing varieties and volumes.

After a few hours on the shop floor, it was fairly obvious to anyone with a lean background that Rowe Cabinets was in trouble with its current operating model. The business had become too large and complex now to rely so heavily on this grocery store of stocked components. Sourcing all the door and drawer parts to inventory wasn't working anymore for the business; it was causing waste through huge door and drawer inventory levels, frequent expedites, and significantly more material handling. And as complexity increased, defects and rework started to multiply too. Rowe Cabinets was spiraling into a lean abyss and needed to change its manufacturing strategies quickly!

What the business specifically needed to do at this point was to develop a make-to-order capability for its doors and drawer faces. The business needed to learn to build more or most of its cabinets completely to order without relying on stocked inventory of configurable components. The business had grown to a point where it had too many product offerings and too much volume to be successful by using a stocked inventory (grocery store) of supplied doors and drawer fronts. The business needed to either create internal lean manufacturing cells that could make the doors and drawer faces to order or develop a system to reliably have these made at outside suppliers to order.

According to Rowe's original seven-day lead time strategy, shared in the chart earlier, Rowe had four days available, days one through four, to manufacture or procure all of these parts and subassemblies and to get them to finishing and final assembly. Four days is plenty of time to execute door and drawer face manufacturing directly to order while maintaining a seven-day overall manufacturing cycle.

Further, with the size of the business now, Rowe might also gain a piece cost savings by vertically integrating its doors and drawer fronts. They might be able to competitively manufacture these components in their own factory and do so in much less space compared to their current build to inventory strategy. Vertical integration was something Rowe Cabinets needed to study and decide strategically if this was something the business wanted to take on. In any case, the business needed to develop a system to make their highly varied cabinet parts to order—cabinet doors, drawer fronts, and drawer boxes.

Rowe could gain back control of its manufacturing performance through this build-to-order capability. Manufacturing operations are faced with the daily challenges of meeting diverse customer orders, volume and mix fluctuations, change orders, reorders, and of gamut of other stresses. The more that Rowe Cabinets can control with its parts and assembly flows through just in time's short manufacturing lead times, the more effective they will be at successfully filling orders. Build to order will give Rowe much better control of its daily execution; it will be able to make only what it needs exactly when it needs it.

Just-in-time manufacturing, either through outside sourcing or vertical integration, is clearly where Rowe

Cabinets needed to head. However, while the need to make this change may have been obvious to experienced lean practitioners, it was not a change the new owners supported. The new owners had bought a successful, working business with the best "lean" model in their corporation. They paid a premium for this high performance business model, and they fully expected the existing, successful lean model to keep printing money the way it had been without a major investment in risky new manufacturing capabilities. Uncoupling the existing door and drawer supply systems in the business and creating a just-in-time capability to manufacture doors and drawers to order was going to be a significant investment and bring significant risks of disruption to the business. This was not a project the corporate headquarters could see that they needed to take on.

Further, the new corporate owners were not steeped in lean manufacturing. The corporation was a traditional batch-and-queue manufacturer without deep lean experience. Shortly after the sale of the business both Bill Rowe and Frank Jackson had departed, so Rowe's lean strategists were gone. And their replacements were not experienced nor credible enough to convince their corporate owners that the business needed to make these critical changes. It can be difficult for corporate leaders to grasp the need to modify their systems that generated past success, even as the signs of serious declining performance start to show. Momentum is a strong force in both the manufacturing and management decision making worlds.

The conclusion of this business case example is that the business didn't change fast enough. Rowe Cabinets stuck to its original manufacturing strategy to the ugly end. They struggled through several years of difficult performance. On-time delivery plummeted, Rowe's costs of quality grew to unacceptable levels because Rowe was now a stressed manufacturing environment, customer satisfaction tanked, and profitability went severely into the red.

After several years of these losses, the corporate owners started looking to sell off Rowe Cabinets to recoup what they could from their investment. The business was a distressed asset at this point and not attractive to potential investors. The recession that started in 2008 was particularly tough on the building products industry, as new housing starts plummeted by more than 80 percent. Rowe's fate was sealed by this recession. No one was interested in purchasing this loss position business in a market that was tanking. As a result, the owners made the difficult decision to close Rowe Cabinets down for good to stop the bleeding.

Within a short ten-year cycle, Rowe Cabinets went from being a very strong business with the best lean manufacturing model in the corporation to closing its doors forever.

Rowe Cabinets is an example of a business that just plain outgrew its early lean strategies. One where the dynamics of corporate governance and decision-making couldn't recognize the changes that needed to take place. Rowe's successful, early lean model of outsourcing key components to inventory became an impediment as the business grew, ultimately causing the business to fail.

Other cabinet companies are using this build-to-order model to succeed through lean manufacturing's short lead times. I've worked with and benchmarked some of these. Rowe Cabinets was simply a case of a lack of lean know how and fortitude to understand and support the changes that the business needed to make to stay successful. We need to continuously re-evaluate our lean strategies as the business evolves and never let the success of where we've been get in the way of where we need to go.

AFTERWORD

I am convinced now more than ever that lean manufacturing will continue to gain ground as the better way to manufacture, as it has for these past two decades. Continuous flow manufacturing is required for the long-term success of our manufacturing businesses.

Avoiding the common pitfalls that come with lean so that we don't end up with a problem instead of a solution takes a focused, proactive, knowledge-based approach to your operations and your organization. The inputs to and outputs from lean are variable. Your success depends on how well you understand what you are trying to do and exactly how to do it. Knowing what you have to change to enable lean will increase your odds of success. It is an overused cliché, but lean is a never-ending journey.

I hope that these lean lessons, along with my overview of the lean manufacturing evolution, will help you be more successful with your manufacturing strategies!

> There is far more opportunity than there is ability.
> —Thomas Edison